EXPLODING
"Conventional Wisdom"

"*Only morality in our actions can give beauty and dignity to our lives*"
Albert Einstein

"*There must be no barriers for freedom of inquiry. There is no place for dogma in science. The scientist is free, and must be free to ask any question, to doubt any assertion, to seek for any evidence to correct any errors.*"
J. Robert Oppenheimer

Boldly confronts… the notion that so-called "conventional wisdom" should be accepted as truth…refreshingly honest in its approach and, by all accounts, is long overdue.
Dr Charlotte Bean – University of Warwick

Once I started to read it, I could not put it down. …The book is replete with wisdom, provokes thought and certainly exemplifies the art of an archivist. I recommend this book without reservation.
*Dr Chris Botton – Centre for Educational Studies,
University of Hull*

This… beautiful book…might make a change towards freedom at last!
*Professor Erik Trell – Faculty of Health Sciences,
University of Linköping, Sweden*

Professor Dunning-Davies is to be thanked for bringing these matters before a broader audience, despite the wails and cries of the magicians. Hopefully, science will one day soon rediscover scientific integrity and be guided by it.
*Stephen J Crothers - Associate Editor,
Progress in Physics (Australia)*

From this most important book there emerges an alarming picture of quite destructive forces abusing scientific power and media positions …and at the same time obstructing recent real scientific advances.
Prof. Stein Johansen - PhD Philosophy of Science, Norway

A long needed addition to the literature, whose goal is to unmask the mafia controlled grip on physics and science in general.
*Prof. Bernard Lavenda - University degli Studi,
Camerina, Italy*

EXPLODING A MYTH
"Conventional Wisdom" or Scientific Truth?

Jeremy Dunning-Davies, BSc, PhD
Department of Physics, University of Hull
Institute for Basic Research, Palm Harbor, Florida, USA

Horwood

HORWOOD PUBLISHING LIMITED
International Publishers in Science and Technology
Coll House, Westergate, Chichester, West Sussex, PO20 3QL.

First published in 2007

COPYRIGHT NOTICE
All Rights Reserved. No part of this publication may be reproduced, stored in a retrieval system, or transmitted in any form or by any means, electronic, mechanical, photocopying, recording, or otherwise, without the permission of Horwood Publishing, Coll House, Westergate, Chichester, West Sussex, PO20 3QL England.

© Horwood Publishing Limited, 2007.

British Library Cataloguing in Publication Data
A catalogue record of this book is available from the British Library.

ISBN: 978-1-904275-30-5

Cover design by Jim Wilkie.
Printed and bound in the UK by Antony Rowe Limited.

Contents

	Foreword	6
	Preface	12
	Introduction	17
1	"Einstein's Biggest Blunder"?	23
2	Einstein's Theories of Relativity	54
3	Big Bang Theory – Controversial or Not?	87
4	The Schwartzschild Solution and Black Holes	130
5	Hadronic Mechanics	181
6	'Conventional Wisdom' Some Modern Case Studies:	
	Black Hole Entropy	222
	The Tsallis Entropy	227
	The Inflationary Scenario	231
	String Theory	232
7	Some Final Thoughts	238
	Epilogue	251
	Index	253

FOREWORD

It is after a considerable amount of soul searching and reflection that I have written this foreword. This delay has been due entirely to my own personal feelings concerning the condition of present day science. It seems to me that we are experiencing the biggest example of scientific obscurantism of modern times. In fact, I feel it dwarfs that experienced by mankind during Galileo's times. The huge difference, however, is that in Galileo's time the major negative role was played by the Vatican for religious reasons, whereas today, the leading role is played by leading organizations and members of the community of scientists; a community now, because of its own internal strife, virtually torn asunder.

In view of the above rather depressing conviction, I have no words to express my appreciation and esteem for Jeremy Dunning-Davies, not only because of his stature as a man of values, courage and conviction, but also because of the value of his scientific work, including, but not limited to, this book. Stated in a nutshell, in the absence of courageous writings such as this book by Professor Jeremy Dunning-Davies, I believe that the current scientific obscurantism will increase, rather than be contained, due to supine acceptance that is implicit in silence.

Professor Dunning-Davies presents numerous cases that can only be regarded as evidence of the said contemporary scientific obscurantism, cases recommended for serious study by serious scholars, not only to implement the necessary corrective measures, but also because they constitute a brilliant list of fundamentally open areas of contemporary science available for investigation by all young minds of any age. The said cases include: a review of the recent saga of a BBC programme concerning the variation of the speed of light, as well as of the stages of the sterile appeal to the British regulatory body; the deplorable scientific condition of Einstein's theories of both special and general relativity, - a field in which nobody is allowed to express a dissident view without expulsion from accepted contemporary scientific society; the background to the 'Big Bang' theory and how the Steady State theory was never properly evaluated; the murky business of Black Holes, pointing out the problems with the so-called Schwartzschild solution to Einstein's field equations, which appears in most books and that no black hole has been identified beyond reasonable doubt simply because Michell's condition, dating from 1784, is not satisfied; an in depth discussion of hadronic mechanics, especially a number of the applications issuing forth from it; stimulating thoughts on four particular cases involving the power of 'conventional wisdom' in black hole entropy, the Tsallis entropy, the Inflationary Scenario and String Theory; then concluding with some, in my view, inspired final thoughts.

On reading the book I formed the distinct impression that Professor Dunning-Davies has been sincerely and deeply worried for years by the pernicious effects of non-scientific influences on scientific research and by the fact that, on many occasions, reasons other than purely scientific ones are used to promote or, more sadly, to reject the publication of professional papers presenting novel theories that are original, well written and plausible, thus deserving publication, as well as the rejection of applications for research funding fully needed by society, yet not aligned with the said organized interests, including the manifestly organized consigning to public oblivion of proven work for reasons which may only be described as questionable. Hence, the raison d'etre for the book, which takes the precise form it does due to Professor Dunning- Davies' own scientific interests which I respect, share and support fully.

During my own active research over a 50 year period, initiated with a paper of 1956 on the ether as a universal medium, written when I was a high school student, I have experienced numerous additional cases in the American, European and Asian continents fully complementing Professor Dunning-Davies' treatment. Among these I would like to indicate as most distressful the discrediting or sheer opposition to the conduction of experiments that might even remotely set limitations on Einsteinian doctrines, quantum mechanics and quantum chemistry, or the sheer manipulation of experiments and/or their data elaboration so as to comply with the said doctrines. Personally, I have reluctantly reached the conviction that, in view of their complexity, contemporary experiments provide suitable prey for manipulations to serve pre-determined goals of the said organized interests.

Mankind is facing environmental problems so serious as to be potentially catastrophic. Here one has in mind the possible halting of the Gulf Stream, after which England will become like Iceland in winter and like the Sahara in summer; environmental problems that have already caused mass exodus of biblical proportions, such as the abandonment by the local Indios of once fertile land at the basis of the Andes due to lack of snow in winter and consequential drought in summer, not to mention the exodus of entire populations in Africa.

The solution to these catastrophic environmental problems can be provided solely by basic advances in scientific knowledge, that is, advances, not in peripheral technological details, but in the very foundational concepts, theories and experiments. Here, to avoid serious problems of scientific ethics and accountability, particularly when public funds are being used, any advance can be considered as being of truly basic character if, and only if, it surpasses pre-existing doctrines.

The identification of all conceivable forms of energies permitted by Einsteinian doctrines, quantum mechanics and quantum chemistry has been long exhausted following research conducted by hundreds of thousands of scientists for about a century. The only hope for mankind to at least initiate the containment of the increasingly cataclysmic climactic changes is that of identifying basically new clean energies that, by conception, are not predicted by the said doctrines, but require their surpassing with broader theories.

The reason for this severe judgment of the contemporary condition of science is that the totality of the large research funds in various countries collectively estimated to exceed ten billion dollars per year, is allocated solely under the condition of full compliance with Einsteinian doctrines, quantum mechanics and quantum chemistry, with a similar occurrence existing in publications at the journals of the American, British, Swedish and other physical societies. Alternative professional research conducted by courageous scientists continues to be opposed, discredited and jeopardized via the abuse of academic power. This is the situation clearly identified and illustrated by Professor Dunning-Davies in this book.

It should be stressed that the above limitations cannot damage the historical value of orthodox doctrines since the said limitations refer to physical conditions not only generally different from those of the original conception, but to conditions often unknown at the time of the said conception. As an example, Einstein's special relativity achieved towering status in the history of science also because of its strict time reversal character, since such a character was mandatory for the representation of the orbits of electrons around nuclei, which orbits are indeed time reversal invariant. But then, scholars seriously interested in serious science, rather than in the pursuit of myopic academic gains, are expected to admit that the very reasons for the success of orthodox doctrines for the conditions of their original conception, are incontrovertible evidence of their limitations for the much broader conditions of contemporary societal needs, thus mandating the support, encouragement and active role in the laborious process of trial and error toward broader theories.

In my view, this short book strikes a real blow for true science. It is to be hoped that scientists will read it with open minds and reflect honestly on what it has to say, that members of scientific organizations from the smallest of national bodies to even the august Nobel committee will also read it in a similar light, and finally that members of the general public, which ultimately funds all our scientific endeavours, will also read it and reflect on the wisdom with which their money is being spent.

Ruggero Maria Santilli
President
The Institute for Basic Research
Palm Harbor, Florida, U.S.A.
September 30, 2006

Preface

There comes a time in the lives of many people when it seems appropriate to reflect on what has gone before, whether it be in someone's personal life or their professional career. In the present case, that reflection has been concerned with some aspects of a professional career after forty years on the staff of a British University, firstly in a mathematics department and latterly, since 2002, in a physics department. Throughout, my research interests have been in the general area of theoretical physics, initially in problems of thermodynamics and, to a somewhat lesser extent, statistical mechanics; more recently, in problems of astrophysics and cosmology. The interest in astrophysics and cosmology has concentrated largely on applications of thermodynamics in these two areas. Hence, the original research interest has persisted throughout. This has nurtured an appreciation for the topic of thermodynamics, as well as an admiration for the work of the founding fathers of that fundamental aspect of theoretical physics and engineering.

I suppose like many on the staffs of British Universities in the latter part of the twentieth century, life initially moved along quite smoothly and cosily, possibly too much so. It was only following a chance meeting at a conference held at Gregynog in North Wales in 1987 that things altered quite dramatically and a new aspect of academia raised its, in this particular case, ugly head. It was shortly after this momentous meeting that the influence and power of **'conventional**

wisdom' in scientific research first became apparent. At the meeting I met Bernard Lavenda and we immediately formed a friendship which has lasted to this day. Shortly after this first meeting, we began considering the validity of the so-called Bekenstein-Hawking expression for the entropy of a black hole. Various aspects of this expression caused us concern from a thermodynamical point of view. Accordingly we wrote a short letter which appeared without any problem in 1988 in the journal *Classical and Quantum Gravity* (**5**, L149). Since it was a letter announcing a new result, we followed it with a full length article which gave precise details of our argument. This article was rejected. No adequate reason for this rejection was advanced and, to this day, nothing has appeared in the said journal pointing out where our original letter was wrong. Although it was not immediately apparent, this incident marked the beginning of publishing problems. Over the intervening years Bernard Lavenda and I have published numerous papers, jointly and separately, on the thermodynamics of black holes but, in all cases, having the articles accepted was rarely straightforward. The same problem occurred in other areas also, such as when we criticised the original theory of inflation due to Guth. **The point to note here is that open scientific discussion was actively prevented by a person, or persons, unknown**; it is not a case of one party arrogantly claiming itself to be definitely correct but rather being prevented from expressing an opinion. Unfortunately, over the years, it became obvious that this attitude was not confined to one or two small areas of physics but to huge swathes of the subject. The entire area of relativity is sacrosanct; it seems that the 'Big Bang' theory for the beginnings of our universe is now regarded as being absolute truth, rather than simply another theory; the Schwartzschild solution of Einstein's field equations is not open for discussion; the existence and properties of black holes are untouchable; the list seems to be endless but these topics are covered in chapters two to four of what follows. More recently, the work of Ruggero Santilli in Florida has been subject to the same problems. He has attempted to extend quantum mechanics, but many of those in positions of real power in science regard the theory of quantum mechanics as complete. This has undoubtedly slowed progress, particularly in examining possible practical outcomes of the new extended theory. True, a new clean energy, 'magnegas', has been produced and is successful, but other possible developments have been prevented – including a possibility for dealing with nuclear waste safely and in a short period of time. These issues are, in my view, far too important to be ignored and are discussed here in chapter five.

The apparent influence on scientific research of what is popularly known as 'conventional wisdom' is discussed here using examples from the general area of physics, although there are links with chemistry when Santilli's work is reviewed. However, the truly disturbing question has to be, if this unhealthy influence exists in physics, will it not exist in other areas of science also? When areas of science such as medicine are considered, this question assumes increasing importance. The possible problems introduced by the constraints imposed by the dictats of 'conventional wisdom' may be compounded when the usefulness and importance of collaboration is added to the discussion. It is known already that collaboration is essential in scientific research except in an extremely small number of cases. This remark refers more to the experimental aspects of the work, rather than the theoretical ones where someone may still come up with a world shattering result via a pencil scribble on the back of an envelope. In general, however, collaboration is the key and should be supported. Obviously this collaboration must be between individuals but also, in many cases, between separate establishments. One establishment must never be so wary of the achievements of another that the two will not share information. However, there are suspicions that this is the case and, if so, one must wonder at the influence of the more prestigious scientific prizes in this. The glory of winning a major prize must never be allowed to play any part in the withholding of collaboration. It may be, and possibly is, completely naïve to think that scientists should be above this sort of totally unscientific influence, but all scientists are only mere mortals. Their work is different but they are subject to the same lures and temptations as everyone else.

This then forms the background to, and the raison d'être for, this book. In the final analysis, members of the general public pay for scientific research in the hope that, eventually, benefits will accrue. It seems only fitting, therefore, that that public should be made aware that the much vaunted scientific ivory tower is not as glitteringly white as it should be.

My personal scientific journey was initiated by Peter Landsberg, who accepted me as a research student when he was a professor at the University College of Wales in Cardiff. It was he who first introduced me to the fascinating subject of thermodynamics – something for which I will be forever grateful. Progress was accelerated by my meeting and subsequent enduring friendship with Bernard Lavenda, surely one of the most unquestioningly supportive of friends. At Hull, George Cole introduced me to astrophysics and cosmology. Our weekly chats over coffee have produced so much more than I am convinced would have appeared after countless hours slaving over books and/or internet references. Finally, Ruggero Santilli. What can I say of him? Like Bernard Lavenda, he is the most steadfast of friends, never withholding his support and advice. This quartet deserves to be recognised for being true open-minded scientists and, along with so many others such as Erik Trell, Stein Johansen and Stephen Crothers, will, I hope, accept my public thanks for all their help and support.

Finally, to my wife, Faith, and children, Jonathan and Bryony, I offer my thanks and undying love, as well as my apologies for boring you (I am sure) for many years with talk of black holes, big bangs, and so much more of somewhat dubious interest to the uninitiated.

Jeremy Dunning-Davies (1st Aug., 2006)

Introduction

In the world of the twenty-first century, science is entering the lives of ordinary people more and more, and in a wide variety of different ways. The most immediately obvious way is through the proliferation of appliances in homes. Not long after the end of the Second World War, television sets began to appear far more commonly in peoples' homes. Some years later, these were joined by video-recorders and players and this has now advanced to DVD players. Gramophones and gramophone records have been superseded by CD players and CD's. No modern kitchen is without a refrigerator and freezer; more and more automatic washing machines, dishwashers and other labour-saving devices are appearing. The garden is no longer a place where heavy manual work is the norm; hand mowers and shears have been replaced by powered mowers – even the sit-on variety is becoming increasingly common – and hedge trimmers. Leaves no longer provide a nuisance in borders with the advent of what may only be described as giant vacuum cleaners for collecting them. Cars have improved in safety and reliability and more automated methods for cleaning them have been developed. All these, and many more, help provide entertainment for people and also reduce the workload of many in our communities. Life is totally different today as compared with even the middle of the last century. Other benefits derived from scientific investigations which directly affect many, include truly stupendous advances in the field of medicine. One obvious example here has to be the work in laser eye surgery, which has progressed so much, so quickly and has meant so much to so many people; people who can see nothing without the assistance of glasses or contact lenses are able, after a relatively short period of time and a little, apparently minor, discomfort, to experience the joy of twenty-twenty vision. Detection methods in diagnosis have also advanced so much in such a relatively short period of time. Medical practitioners can now call on the help of a wide variety of scanning processes, not just rely on the use of X-rays. The delivery of doses of radiation is now being investigated as a truly serious theoretical problem; no longer is the dose, or its method of delivery, a mere guess. The whole list is virtually endless, but what has been mentioned here serves to illustrate just how important science – both pure and applied - and scientific advances are in everyday life.

The importance of the appliances mentioned and, indeed, of many others in everyday life, together with the increasingly intrusive nature of the media, has brought science in general more to the attention of everyone. All branches of the media are heavily involved in advertising. Information about new and/or improved products seemingly assails us from all directions. Lengthy articles and television programmes make everyone fully aware of new developments in all areas of science – sometimes sooner than it is really sensible to make that information readily available to the general public. Sometimes also, the slant of a television or radio science programme may be such as to convince the members of the general public that some particular theory is, in fact, an almost unassailable truth, rather than just a mere theory. If this does occur, and it is not unknown for it to happen, the question of who initiated the misleading emphasis is raised. This notion, in turn, leads to the possibility of the media being manipulated for the purpose of interested parties. It must be realised that the old adage of 'there is no such thing as bad publicity' applies to scientific publicity as much as to any other area. Exposure of a scientist and his theories in a prime time television programme will obviously increase that person's public standing but, more importantly, will bring the work to the attention of many more people and, in particular, to the attention of those controlling research funds of one form or another. This may seem to imply a very cynical view of the world of scientific research – far removed from the idealistic 'ivory tower' imagined by many – but how far removed is it from reality? Indications of a possible partial answer to that question will emerge, hopefully, in what follows. In the final analysis, however, it will be for the individual reader to reach his/her own conclusion, always remembering that, through the public financing of research, all are making a contribution and so all might feel entitled to a say in how the distribution is achieved.

Usually, it seems that television producers take a lead from what is currently in fashion in the pages of some of the semi-popular science journals – those journals which, although purporting to be high-brow science journals, retain a semi-popular image and are readily available on news-stands. Careful perusal of these journals indicates that, at any one time, certain topics are definitely 'in vogue' while others are not. On top of this, there are some theories which are always 'in vogue' and are regarded as being virtually sacrosanct; no criticism, either implied or direct, of these is allowed, for some unfathomable reason. This is the realm of 'conventional wisdom'; that body of knowledge – in this case scientific knowledge – which is accepted by those in positions of power to be unchallengeable. It is the purpose of what follows to examine several areas of perceived 'conventional wisdom' in physics; to look at the physics involved and note the implications for future scientific advance. Much of the discussion, by its very nature, has to be speculative, but the possible effects for mankind are there for all to see. When it comes to considering the relatively new field of hadronic mechanics, which will be discussed in more detail in chapter five, it immediately becomes all too apparent that although much of the work completed has been theoretical, the possibilities for mankind are immense if experiments and observation support that theory. However, hadronic mechanics does not agree with some aspects of perceived 'conventional wisdom'. Also, it could be seen as posing a threat to some areas of big business – particularly that highly lucrative business concerned with the transportation and disposal of nuclear waste - if proved correct. Hence, it is possibly not too surprising to find that great difficulty is being experienced by the theory's originator and his collaborators in persuading someone to search for the required experimental support; quite simply, so far, no-one is willing to perform the necessary experiments. The cost of these experiments would not be excessive and so, even if the theory was proved incorrect, not too much would have been lost financially. On the other hand, as will be seen later, if the experiments proved the theory correct, the benefits for mankind in the fields of energy production and disposal of nuclear waste would be out of all proportion to the cost.

However, it does seem that 'conventional wisdom' is in a position of such power as to be virtually unassailable, although that position must ultimately be totally stifling for true scientific advance. Here a quite specific incident will be reviewed in the first chapter before going on to consider the situation relating to certain general areas of physics; questions concerning the theories of relativity, the 'Big Bang' and the modern conception of black holes will be considered, before looking at the ideas and possible consequences of hadronic mechanics in a little more detail. Finally, attention will return, albeit briefly, to further specific situations before some final conclusions are suggested for consideration. The situation is, in the view of many, serious, and any one person's real knowledge only covers some fields of human endeavour – in this case, some areas of physics. If, as is widely suspected, the problem covers all areas of scientific endeavour – including medicine – then it is indeed a problem which demands an immediate resolution for the benefit of all – scientist and non-scientist alike.

Chapter One

"Einstein's Biggest Blunder"?

This was the title of a rather interesting and well-made programme which appeared as an edition of *Equinox* on Channel 4 of British television on 23rd. October, 2000. The programme was quite wide-ranging in the topics it covered, but the thing that stood out was that the two researchers, who were the main contributors, were discussing the possibility of the speed of light varying, seemingly in direct contradiction to Einstein's theory of special relativity, which regards the speed of light in a vacuum to be the same for all observers, and using that to attempt to explain away some of the problems facing the generally accepted Big Bang theory. All this appeared to be being achieved within the theory of general relativity - a theory which was developed by Einstein from his earlier special theory of relativity and is really a theory of gravitation. This latter point was not explored although, considering the fact that the basis of special relativity is the constancy of the speed of light in vacuo, it is difficult to appreciate how a theory involving a variable speed of light can remain within the confines of special relativity. Be that as it may, the detailed content of the programme could have caused either upset or, indeed, offence to a number of people due to the fact that the whole idea of introducing a variable speed of light

Exploding A Myth

appeared to be being claimed as an original idea by the two researchers involved, Andy Albrecht and Joao Magueijo, who were both working at Imperial College, London, at the time. In fact, Magueijo actually claimed that "we did something which most people consider to be a bit of heresy. We decided that the speed of light could change in space and time, and if that is true then our perceptions of physics will change dramatically."[*]

The remainder of that section at the very beginning of the programme speaks volumes also. The narrator proceeded to comment that "at the dawn of a new century, a new theory is being born. It threatens to demolish the foundations of twentieth century physics. Its authors are two of the world's leading cosmologists. If they're right, Einstein was wrong. It all began when Andy Albrecht and Joao Magueijo met at a conference in America in 1996."[†] Albrecht then continues, claiming that it "was pretty exciting. Most of the key people were there and there were lots of debates about the contemporary issues in cosmology. Joao came up to me one evening and had a very interesting idea"[‡]. Magueijo then takes up the story, saying, "This is total bullshit! It wasn't like that at all. I remember there was this conversation between the three of us, and each one of us suggested something. I remember I suggested the varying speed of light and there was an embarrassed silence. I think you two thought I was taking the piss at this point."[§] Magueijo continues "But then, oh he's

---
[*] Transcript of the aforementioned mentioned television programme
[†] Ibid
[‡] Ibid
[§] Ibid

Einstein's Biggest Blunder?

actually serious, he's not laughing; then we started taking it more seriously."* One wonders if this conversation tells more of Magueijo himself than of the scientific story. Again, if they were indeed regarded as 'two of the world's leading cosmologists' at the time, how many leading cosmologists were there then? They must have been two out of a very long list; a list so long as to stretch beyond credibility the usual meaning of the word 'leading'.

The programme itself continued by noting that "for most scientists the idea that the speed of light can change is outrageous; it flatly contradicts Einstein's theories of space and time"[†]. It was then pointed out that, more recently, people had begun to question whether or not the universe itself actually behaved as Einstein's theories would have you believe. It was mentioned that measurements have been made which indicate that the universe seems to be filled with a kind of energy density which is not understood and that the expansion of the universe seems to be speeding up, rather than slowing down. This was felt to indicate that the laws of nature were not always as they are now and this appeared to indicate an impending revolution in physics for the beginning of the twenty-first century comparable with the one at the beginning of the twentieth century. This was a fascinating opening and still seems so in retrospect, but it certainly assigned to Magueijo a position of scientific originality which is difficult to justify; but more of that later. The

---
* Ibid
† Ibid

programme continued with an examination of the history surrounding some of Einstein's original papers, how physics had developed as a result, and eventually began to grapple with the problems remaining with the Big Bang, in particular inflation. Albrecht talked about the whole idea of inflation and how, although he had helped develop that particular theory, he had never been totally convinced that it was the correct solution to the problems facing the Big Bang. Apparently, Magueijo had also harboured doubts about inflation. It was claimed that Magueijo realised that, if you broke "one single, but sacred, rule of the game, the constancy of the speed of light"[*], you could solve the big outstanding problem. However, he admitted that, at the time when he felt this, it would not have been politic to pursue this idea. This is very true and tells us something very basic about scientific research; that is, you have to obey the unwritten rules and never ever go against perceived 'conventional wisdom'! This perceived 'conventional wisdom' is undoubtedly more important than scientific truth; indeed, it seems that, only too often, the pursuit of scientific truth can be to the detriment of a person's career - particularly if that pursuit involves challenging so-called 'conventional wisdom'. Magueijo, therefore, kept his thoughts to himself until he was awarded a research fellowship and joined Albrecht's group at Imperial College. The two then worked on the idea behind closed doors, apparently even cleaning the blackboard after each session. The culmination of this work was an article published in the prestigious journal,

---

[*] Ibid

Einstein's Biggest Blunder?

Physical Review D[*]. These final two points undoubtedly form an uneasy alliance, given the usual attitudes of modern science. On the one hand, the researchers seem highly conscious of the fact that their thoughts are leading them to oppose 'conventional wisdom'; on the other, their written up results appear as an article in a journal known not to oppose 'conventional wisdom'. For many, it will be something of a puzzle as to how this article ever appeared in such a journal and the question of who sanctioned its publication, and why, will be uppermost in many people's minds – even if the thought remains unspoken.

Some quite outstanding claims on behalf of the two researchers were then made in the continuation of the programme. They were, for example, credited with "creating a completely new physics". The conclusion, however, is what really raises further important questions. Magueijo finally claimed that he respected 'relativity enormously' but had the feeling that it was only now that he had contradicted relativity that he really understood it. He went on to state that it was because he'd gone against relativity that he was showing his "full respect to the great man" (Einstein). He was not contradicting Einstein, merely attempting to take things one step further. He is sensible enough to acknowledge that eventually it will be nature that decides which is the correct explanation of things. The whole story does, however, raise a number of important issues concerning scientific research and some relate

---

[*] A. Albrecht and J. Magueijo, 1999, *Phys. Rev. D*, **59,** 043516

Exploding A Myth

directly to why the programme could have caused both offence and upset to some people.

The claimed 'revolutionary' theory of Albrecht and Magueijo eventually appeared in an article in *Physical Review D* in 1999. By then, objections to its total originality had already been raised by John Moffat in Canada and, as a result, a note was added, apparently at the proof stage, acknowledging this previously unmentioned work. Moffat had, in fact, published two articles dealing with a variable speed of light in 1993[*] However, this 'new' theory probably received its biggest public boost via the programme produced for Channel 4 of British television, by Dox Productions, under the heading of *Einstein's Biggest Blunder*, which, as mentioned earlier, appeared as an edition of Equinox on Monday, October 23rd, 2000. The idea of a varying speed of light goes back a long way and has certainly been the subject of speculation ever since Einstein stipulated a constant speed of light in vacuo as a basic assumption behind the special theory of relativity. It is illuminating to note that Einstein himself refers to the speed of light varying with the gravitational field in an article of 1911 entitled *On the Influence of Gravitation on the Propagation of Light*, which appeared in the prestigious German journal *Annalen der Physik* and today is reproduced in the book *The Principle of Relativity*[†]. There is also reference made to this in another article by Einstein, which also appeared in

---

[*] J. Moffat, 1993, *Int. J. Modern Phys. D,* **2**, 351, 1993, *Found. Phys.* **23**, 411

[†] A. Einstein, 1911, *Ann.Phys.*, **35**, 898 (Also see '*The Principle of Relativity*', Dover, 1952)

Einstein's Biggest Blunder?

Annalen der Physik, in 1912. It could be argued, not unreasonably, that such publications do not, however, make any reference to such an idea having any potential for solving cosmological problems. It might be argued also that Einstein's paper of 1915 superseded these earlier works because they could be considered preliminary considerations which helped towards the final formulation of the world shattering Theory of General Relativity. However, that same argument does not hold for more recent articles such as those of Moffat or, more particularly, those of Thornhill which appeared in 1985[*]. The crucial thing about the Thornhill articles, one in particular, is that, as well as predicting the speed of light to vary with the square root of the temperature, all the details of his work, together with the comment that, if it were true, it would do away with the necessity for a theory of inflation, were reported at an International Conference entitled *Physical Interpretations of Relativity Theory*, and held at Imperial College in London in September 1996. The contents of this lecture were published also[†].

However, to continue the saga of the television programme. At the suggestion of Channel 4, who obviously wished to have no more to do with the matter, Dox Productions was contacted. The whole object of contacting both Channel 4 and, subsequently, Dox Productions was to obtain recognition of the earlier relevant work of Thornhill. No-one was attempting to

---

[*] C. K. Thornhill, 1983, *Speculations Sci. Tech.*, **8**, 263

[†] G. H. A. Cole & J. Dunning-Davies, 2001, in *Recent Advances in Relativity Theory*, vol. 2 (eds. M. C. Duffy & M. Wegener), 51

Exploding A Myth

discredit Albrecht and Magueijo, but they had been beaten to publication on this important topic and by quite a few years. The claim made about the actual programme was simply that it was not researched well enough and it was requested that the situation be clarified with credit being given where credit was undoubtedly due. In some ways surprisingly, the programme makers felt totally justified with their coverage. They pointed out that they made "no claims whatsoever for Dr. Magueijo's priority in this matter". Further they stated that "the point is not that Joao Magueijo and Andy Albrecht were the first people to suggest that the speed of light varies, they were not. However, they were the first scientists to work out many of the implications of this in the context of modern cosmological ideas." Even this is not entirely true since, as mentioned earlier, the effect of a varying speed of light on inflation had had attention drawn to it already, and at an International Conference. Also, the precise functional relationship between the speed of light and temperature was the real issue, rather than the fact that the speed of light was not a constant. In fairness, the claim that "a television documentary is not a scholarly article" and "cannot by its very nature include the scholarly apparatus of footnotes and references" is reasonable. It is also true to say that, "while establishing priority is something that scientists may care about, it is not something that necessarily aids the understanding of an idea." Again though, it is surely not unreasonable to hope that the person originally responsible for an idea should receive some credit? It is certainly reasonable to hope that, if something is

Einstein's Biggest Blunder?

omitted inadvertently, that omission will be rectified if, and when, attention is drawn to it. It was even agreed with Dox Productions that, although in the transcript of the Equinox programme it appeared that Albrecht and Magueijo were claiming total originality for the work, those at Dox could not have been expected to know all this - but Albrecht and Magueijo should have. The end result appeared to be that the programme ascribed far more credit to Albrecht and Magueijo than they could claim reasonably. That being so, it seemed that the balance should be restored, with credit being given where credit was due.

With no sign of movement by the broadcasters, the whole affair was referred to the Broadcasting Standards Commission which, after collecting details, was impressed enough by the seriousness of the situation to convene a hearing in London in the July of 2001. The adjudicating panel at this meeting was composed entirely of lay people. In the event, opening statements were made by both sides, a certain amount of cross questioning followed, and the meeting concluded with closing statements by both sides. The programme producers had brought along Professor David Wark from the University of Sussex, who had also appeared in the programme under discussion, as an expert witness. His evidence was entirely technical and he continued with this approach even after the Chairman specifically asked that no highly technical discussion take place as the panel was composed entirely of scientifically lay persons. The main argument advanced against acknowledgement of Thornhill's work was that

Exploding A Myth

he had published in obscure journals and it was unreasonable to expect anyone to have found these references. It was pointed out, quite legitimately, that such an excuse for exclusion of a reference is not acceptable in British Universities, even for undergraduate project work; indeed, if an undergraduate committed such an offence in a project, he could be accused of plagiarism and, if found guilty, sent down from university permanently. It is also of interest to note that Albrecht and Magueijo made a similar claim initially for not noting Moffat's work. However, Moffat's two relevant papers appeared in the *International Journal of Modern Physics* and *Foundations of Physics*. It is undoubtedly the case that such a claim cannot be made as far as these two journals are concerned. Thornhill's articles, on the other hand appeared in the *Proceedings of the Royal Society* (although in this case the title may have been a little misleading as far as a variable speed of light is concerned, but then exactly the same comment could be levelled at Moffat's relevant articles) and in *Speculations in Science and Technology*. This latter journal is a little obscure but, in these days of efficient computer searches, it is difficult to believe the reference was unobtainable. Further, since Thornhill is a regular attender at the meetings of the Royal Astronomical Society, it is unlikely that all of Magueijo's acquaintances would have been totally unaware of his views and the fact that much of his work was published, albeit not in mainline journals! This latter point is particularly relevant since, in his recently published book, *Faster Than the Speed of Light*, Magueijo makes great play of his meetings with John Barrow at the Royal Astronomical Society.

People may speculate, not unreasonably, on why Thornhill's work rarely appeared in the mainline journals if it was apparently so important. This brings up the whole question of scientific research and so-called 'conventional wisdom'. In scientific circles, it is decreed, for example, that Einstein's theory of special relativity is sacrosanct; it is indeed part of perceived 'conventional wisdom'. Therefore, anyone submitting an article to a front line journal which even appears to question the validity of special relativity will almost certainly have that article rejected for publication. Indeed, if a paper in which the word 'aether' appears is submitted to a front line journal, it stands an excellent chance of immediate rejection. In such cases, people must either forget their own work or publish in less well known journals. This is the case with Thornhill's work which is deemed, quite correctly, to be sceptical about the validity of special relativity. This whole scenario raises once again the intriguing question of how the paper by Albrecht and Magueijo was ever allowed to appear in such an apparently establishment journal as *Physical Review D*? The basic topic of the said article undoubtedly brings the theory of special relativity into question, as Magueijo himself admits, so the questions of why it was published and who sanctioned that publication are intriguing for speculation.

As for the complaint to the Broadcasting Standards Commission, it was not upheld. The Commission noted that the complaint was intended as constructive criticism but took the view that "such originality as was claimed in the programme was not for the VSL (varying speed of light) theory itself, but for the work of Dr. Magueijo and Professor Albrecht in their particular field of cosmology". The Commission was "persuaded that Dr.Thornhill's work was significantly different and that there was therefore no obligation on the programme-makers to include any reference to it". The Commission, therefore, found "no unfairness to Dr. Thornhill". The possibility of an appeal against this adjudication existed but, when information on how to proceed with such an appeal was requested, no reply was received from the Commission. However, it is extremely doubtful that proceeding with an appeal would have been worthwhile. In the end, although the Chairman requested that no detailed scientific discussion take place, it was the somewhat jumbled and indeed discredited scientific discussion which apparently won the day. The basic argument about the availability of Thornhill's articles, due to their appearance in obscure journals, was shown to be entirely false but, ultimately, for some unfathomable reason, this was not accepted. Also, it is extremely interesting to note that "The Commission is persuaded that Dr. Thornhill's work was sufficiently different and that there was, therefore, no obligation on the programme-makers to include any reference to it" because this line of reasoning was never specifically discussed; it is impossible, therefore, to even imagine how the Commission was thus persuaded. The crucial point is that, although Dr. Thornhill certainly didn't go on to discuss cosmological implications of his result, the vitally important starting point for Albrecht and Magueijo's cosmological discussions was the speed of light varying, and varying with the square root of the background temperature; without that result, they had no starting point, nothing! That starting point was the result derived by Thornhill many years earlier! Sufficiently different? It is impossible to see where! It might be noted, however, that the crucial point about Thornhill's work is that, for him, the speed of light varied as the square root of the background temperature; there may have been many other theories before him which considered a varying speed of light but the question is did they consider a speed of light with such a temperature variation? If so, they are deserving of recognition also!

The end result of all this was for a piece of scientific work to gain a degree of recognition because of a television programme. Although published in *Physical Review D*, the Albrecht/Magueijo article had not achieved enormously widespread publicity before this programme. That is not to say that it was unknown before the programme, but knowledge of its existence was certainly not as widespread as might have been expected for such a seemingly 'revolutionary' theory. This further thought raises even more speculation about the appearance of such an article in such a prestigious establishment journal, which normally seems to guard 'conventional wisdom' with such vigour. It might be remembered that the original paper on the theory of inflation by Guth appeared in *Physical Review D*[*] but,

---

[*] A. Guth, 1981, *Phys. Rev. D*, **23**, 347

when this paper was shown to be in error thermodynamically, *Physical Review D* would have nothing to do with the correction - after all, by that time, inflation had become part of perceived 'conventional wisdom' and a great number of articles on inflation had already appeared in *Physical Review D*; the journal couldn't be expected to publish a short article showing where a large number of previous papers was in error, or could it? If scientific research is truly what the public is led to believe it to be, then surely any reputable scientific journal should be only too willing to publish anything which advances human knowledge, even if that meant tacitly admitting that many earlier articles contained errors which had been missed by the journal's referees? Considering the above discussion, it is a little ironic that the article pointing out the error in Guth's paper finally appeared in *Foundations of Physics Letters*[*], a sister journal of *Foundations of Physics*. No doubt the editor of those two well-known journals will be delighted to hear that many regard them as obscure! It is interesting to note, considering he now claims to have harboured doubts about the correctness of inflation almost from the outset, that Albrecht doesn't appear to have acknowledged publicly the fact that Guth's original article has been shown to be incorrect. Maybe this also reveals something about human nature and, possibly more particularly, about the power and attitude of the ruling 'mafia' of world science.

---

[*] B. H. Lavenda & J. Dunning-Davies, 1992, *Found. Phys. Lett.* **5**, 191

Einstein's Biggest Blunder?

Another possible insight into the attitudes pervading the establishment, or at least establishment supporting figures, of world science is supplied by the rather childish display of Professor David Wark in relation to the above mentioned complaint to the Broadcasting Standards Commission. He was asked to provide a written submission to the Commission, possibly because he had been a contributor, albeit a minor one, to the original programme. Actually, given that it was a British made programme, the question of why a non-Britisher was imported to make this contribution springs to mind immediately, particularly since the material covered by the actual contribution was not of so high an intellectual level. Was this a factor introduced merely for a possible American market? However, in his written statement, Professor Wark resorted immediately, and for no apparent reason, to purely personal abuse. This fact was never mentioned in any report and he was never rebuked for it at the Commission hearing, but that is what part of his written submission amounted to - pure childish personal abuse!

Professor Wark began by misunderstanding the details of the complaint. Basically the complaint drew attention to the undeniable fact that Thornhill's work predated that of Albrecht and Magueijo; it made no mention of any work by Dunning-Davies, merely mentioning that he had drawn attention to Thornhill's work at a conference held at Imperial College. Professor Wark continued by pointing out that he had "been active in the field of astroparticle physics, and its implications for cosmology, for twenty years", but had never heard of either Thornhill or Dunning-Davies. This is a fair comment but not a surprising revelation since neither person mentioned works specifically and solely in that field. He continued by stating that he had never heard of the *Hadronic Journal* or *Speculations in Science and Technology*, two journals in which Thornhill had published. Again, this is not too surprising a claim, but indicative of the attitude of this person. Although, being an American theoretical physicist, it might not have seemed too far fetched to think that Professor Wark might have heard of the Institute of Basic Research in Florida and the journals, such as the *Hadronic Journal*, it publishes. As explained earlier, if someone wishes to challenge the rigidly held views of the establishment, many big name journals are excluded from consideration. It is for this precise reason that Albrecht and Magueijo should have checked through the less well known journals in their literature search - if, in fact, they ever carried out a literature search at all. It might be remarked again, at this point, that Moffat's articles appeared in well-known journals, although the titles of the relevant articles might have been a little misleading as far as variable speed of light was concerned. It seems, however, that, as far as Albrecht and Magueijo are concerned, any and every argument is allowable as an excuse for their omitting references to already published work; - either the journal is too obscure (whether it is or isn't seems immaterial) or the title was misleading so they couldn't have been expected to realise that a variable speed of light was under discussion. These excuses are not totally unreasonable but it seems odd that there has been so much establishment defence of this decidedly non-establishment piece of work.

Professor Wark then continued by noting that Albrecht and Magueijo had published in *Physical Review D* and that "a search of SPIRES (a standard bibliometric tool in particle physics to find articles and citations) found only 2 citations to the entire work of Dr. Dunning-Davies, while there were 46 citations to just one of the papers by Albrecht and Magueijo (a benchmark can be supplied by the fact that papers need 50 citations to enter the "most-cited lists.)" This sort of pathetic quote says much about Professor Wark and, by implication, the establishment he appears to serve so slavishly. It is hardly surprising that he finds few references to the work of Dunning-Davies in a particle physics bibliometric tool, as he works mainly in other areas, as does Thornhill for that matter. It is, however, surprising that SPIRES appears not to include the highly prestigious journal, *Proceedings of the Royal Society of London* in its listings. As for large numbers of citations of a particular paper, that can occur in several ways. True, it can occur because a particular article is a landmark publication, but such articles are few and far between. However, it can occur also when every member of a research team or group quotes the paper. How does one distinguish between these two possibilities? In all honesty, except in the case of the very occasional outstanding breakthrough article, one cannot. Placing too much reliance on these citation claims is a dangerous and possibly misleading practice, as is the case whenever total reliance is placed on a list; - this practice has been found misleading in so many areas, not just in the evaluation of the scientific merits, or otherwise, of an academic article.

There is one further point to which attention must be drawn concerning the testimony of Professor Wark. He claims, in his submission, that Dr. Dunning-Davies apparently states, 'Thus relativity and the universal constancy of the wave speed of light were entirely discredited in the published literature long before Dr. Joao Magueijo and Professor Andy Albrecht claim to have 'decided (!) that the speed of light could change in space and time'. Apart from the offensive personal comments alluded to earlier, this indicates a slovenly approach to producing a requested statement. Although the above statement was made in a submission to the Commission, it was not made by Dr. Dunning-Davies, as was *very* clearly stated in the mentioned submission.

These two incursions by Professor Wark do little other than tarnish the reputation of scientists and show clearly that the world of academia is not as pure as many in the general public would imagine. It is not a coming together of scholars in an attempt to discuss and, hopefully, solve some of the problems facing the world and individual societies; it is rather a gathering of a wide variety of people, some interested in science from a purely intellectual point of view, others (possibly the majority) using science as a means to further their own personal ambitions. In the latter category, it soon becomes clear that, in the pursuit of their personal goals, anything is allowable provided you

Einstein's Biggest Blunder?

can get away with it; - it is a totally no-holds-barred jungle! This, unfortunately, seems to pervade all branches of science - probably all branches of academia - and this could prove truly tragic in some fields. It is always worth remembering that C.P.Snow's writings came out of his own experiences. This does not mean that he was recording actual events in such as *'The Affair'*, but the sentiments and attitudes expressed in such works surely mirror events witnessed in real life!

Recently, the issue of *Einstein's Biggest Blunder* has been revived with the publication of the book *Faster Than the Speed of Light* [*] by Joao Magueijo, which was mentioned earlier and in which he discusses what he terms "the story of a scientific speculation". It is of particular interest, because of what has gone before here, to note that, on page 123 of this book, Magueijo comments that "it is sad that all too often credit is given not to the people who conceived a new theory but to those who come afterwards and clean up the fine details". This is only too true and is usually regarded as being a case quite simply of 'that's life'. However, it does raise the question of why Magueijo and Albrecht were defended so vigorously against suggestions that they should have quoted Thornhill's work. It is even more surprising when the overall tenor of the discussion in his book seems to indicate that he and Albrecht were so wrapped up in their thoughts about a varying speed of light being a possible way out of the problems associated with the Big Bang, that they weren't really bothering to check through for possible references to

---

[*] J. Magueijo, 2003, *Faster than the speed of light*, (Perseus Books, U.S.A.)

earlier theories involving a variable speed of light. Fundamentally, this is why Moffat's work was probably not referenced initially. It is not totally unreasonable if someone is so involved with a piece of work that, at least initially, they fail to check to see if someone has already had a similar idea. When the time comes to write up, though, then a full literature search should be conducted. Even so, in these days of so many journals and so many articles, it is understandable if references are missed but then, when the omission is pointed out, you must acknowledge your error and do so with good grace; any other reaction must be deemed unacceptable. Indeed, again as mentioned earlier, in many British universities, if an undergraduate failed to give due acknowledgement to published information in any piece of assessed work, that individual could face a lot of trouble. In an extreme case, they could be sent down from university for plagiarism but, in lesser circumstances, they could face unpleasant disciplinary procedures which could result in marks being reduced, even to zero, and having a black mark indelibly attached to their file. This, although true, seems very harsh, particularly when it should be realised that undergraduates have far less experience and knowledge than professional researchers when it comes to checking out references. Of course, with any undergraduate, the ultimate responsibility for ensuring that a complete list of references is attached to any piece of assessed work should rest with the supervisor but, unfortunately, that is not always the case.

Einstein's Biggest Blunder?

The actual book by Magueijo evokes a wide range of reactions and raises anew various questions referred to earlier. It is of particular interest to note that, on page 265, under the heading of acknowledgements, he points out that he is 'above all indebted to David Sington, the producer/director of the documentary *Einstein's Biggest Blunder*, who showed me the way to this book.' Incidentally, it was David Sington who also pushed very vigorously to defend Magueijo and Albrecht over their omission of any reference to the work of Thornhill concerning a variable speed of light. However, as for the above quote from the book, it is obviously not completely clear what is meant by the latter part; - "showed me the way to this book", - what precisely does that mean? Also, what possible interest could a television producer/director have in the production of such a book? It raises, once again, the dominating question of why the 'establishment' seems so determined to promote this person and his work. When read, the book is full of claims about new theories of variable speed of light - note that since the initial programme, it has become theories, rather than just theory - but, due to the very nature of the book, there is very little by way of substance to support these. Given the detailed content, it also raises the question of who is Magueijo's mentor? Throughout the book, he is gratuitously rude and offensive to so many, in particular as well as in general, that it is almost inconceivable, considering the way things are run in both British and World science, that he does not have an extremely powerful protector. But why? Here is a young man who has enjoyed all the benefits of a good education - a

Exploding A Myth

postgraduate at Cambridge - which almost appears at times something he regrets, recipient of, in his own words, a prestigious Royal Society research fellowship (who pushed for him to receive that?), research at Imperial College followed by a Readership at Imperial. What a line-up! After all that, he writes a book in which he attacks so many it is almost laughable, but does so using language more appropriate to the gutter than an apparently well-educated person. He may have wandered around somewhere like the centre of Hull on a Saturday afternoon and been deceived into assuming that four-letter words are a necessity for the ordinary 'man in the street', but he should have enough intelligence to realise that that is not so. This common language is offensive to many, as is evidenced incidentally by some of the reviews printed on the American Amazon web page, and really detracts from the book, - always assuming, of course, that the book does contain material of actual merit.

Actually, the reviews of this book appearing on the American Amazon web page provide illuminating reading; the editorial reviews, in particular, making one wonder from where all these 'facts' have come. "Jocular, ironic, witty, self-centred, even indignant", all terms used to describe Magueijo who, "in spite of his own stature within learned gates", sounds embittered. The colourful language is impressive but is it justified? Magueijo is also described as "a natural teacher", "a fine scientist", "a brilliant scientist", but is he any of these? Many of these remarks seem purely for publicity and might surely be described as a little 'over-the-top'.

Einstein's Biggest Blunder?

Only time will tell how good a scientist and teacher Magueijo is or becomes. Incidentally, the editorial reviews contained little condemnation of his totally unnecessary use of so-called Anglo-Saxon vocabulary. It is extremely doubtful that such vocabulary is ever acceptable in decent society, so maybe this lack of condemnation says more about the writer than the work under review. Some of the customer reviews, however, readily condemn his language, but too many are taken in by his apparent erudition - much in the same way that many are deceived by the so-called popular writings of Hawking. This, of course, is the real danger. When read by laymen, such books can appear much more important and, indeed, powerful than they really are. The authors can achieve for themselves a status which would not be naturally accorded them by their scientific achievements. This can easily lead to a status in popular culture influencing the person's status within the scientific community, simply because everyone has heard of that particular person. A typical case of this is provided by Hawking's book *A Brief History of Time*[*]. This book is undoubtedly a massive best-seller, but why? It is not particularly well-written, contains some indifferent English, and doesn't answer questions it poses early on. As a 'popular' science book, it compares unfavourably with those by Paul Davies and, as far as the actual topic covered is concerned, is both inferior to, and more expensive than, Jayant Narlikar's *The Primæval Universe*[†]. Obviously, the latter book indicates that having a 'big name' in science write a

---

[*] S. W. Hawking, 1988, *A Brief History of Time*, (Bantam Press, London)
[†] J. V. Narlikar, 1988, *The Primaeval Universe*, (Oxford U. P., Oxford)

good popular scientific book is no guarantee of publishing success. So what was, and is, so special about Hawking, since his area of interest is highly abstract and not immediately accessible to the layman? One not unreasonable explanation is that, for some reason, Hawking has become almost a cult figure. Is someone attempting to turn Magueijo into another cult figure? If so, why? Here we have another member of the scientific fraternity who, when you look beneath the surface, doesn't appear too pleasant an individual. Rather than jocular and witty, he would seem to be arrogant, self-opinionated and unnecessarily rude! Many will have much sympathy with some of the sentiments he expresses, but would dissociate themselves from the manner of their expression. This would be particularly true of his views of the actions of the editors of many highly regarded scientific journals, including *Physical Review D*. It would, incidentally, be very true also of at least one of the ex-editors of the apparently well respected journal *Classical and Quantum Gravity*. In this latter case, the ex-editor is now a professor at Imperial College but, when challenged over a totally ludicrous editorial decision, took refuge behind a refusal to enter into any further correspondence on the subject. This is the normal response in such situations and was, no doubt, the sort of response John Moffat received when he submitted his original paper on varying speed of light to *Physical Review D*. He would have known further argument was a complete waste of time and energy. It makes one wonder just how Albrecht and Magueijo were able to even pursue a two-way correspondence with *Physical*

Einstein's Biggest Blunder?

Review D for so long; the fact that they finally achieved success in having their article accepted for publication would be seen by many as a secondary issue. How did they manage to maintain the claimed lengthy two-way correspondence? It might be remembered, for example, that refusal to publish an article which might possibly rock the boat for the perceived 'conventional wisdom' on the current ideas surrounding the entropy of black holes seemingly has to be applauded by all. That would possibly be the justification advanced by one of Magueijo's colleagues at Imperial College for his actions those few years ago.

As far as these so-called prestigious journals are concerned, it must first be noted why they are deemed 'prestigious'. Many have a truly enviable history and have been responsible, in the past, for publishing many of the boundary shifting articles which have resulted in our present state of scientific knowledge. However, most of these journals, probably because of their heritage, are now part of the worldwide scientific establishment and, as such, have become devoted to protecting 'conventional wisdom' against all intrusion. In recent years, the one journal whose editor Magueijo singles out for special abuse, *Nature*, has certainly found itself in this category. It may, in the past, have published some truly landmark papers but its more recent record is, to some, not so noteworthy. However, its editor and its editorial decisions remain unchallengeable usually and this stance seems to be supported by its publisher who appears unwilling to interfere. In fact, some years ago, the position had deteriorated so much that, when an editorial appeared which contained several schoolboy level errors, the journal was extremely dilatory over a letter correcting these. After a considerable period of time had elapsed, the entire matter was referred to the Press Complaints Commission which rapidly ordered the editor to publish the said letter. It is surely a sad reflection on the editors of well-respected scientific journals that such an action should ever prove necessary in order to publicise scientific truth! It is possibly of even more interest to note that, more recently, when a similar situation arose with *Nature* - although this time it was a Letter to the Editor involved rather than an editorial - the Press Complaints Commission ruled against the complaint. In this latter case, the letter concerned contained material which was manifestly incorrect and so the unfortunate ruling by the Press Complaints Commission, a ruling which goes directly against one of its own precedents, has effectively given carte blanche to editors of scientific journals in Britain to do exactly as they personally like. The original precedent of making an editor publish a correction to an error in an existing article could conceivably be open to abuse, but this newly created precedent is truly opening a Pandora's Box. If it is followed through to its logical conclusion, it is difficult to imagine a return to total honesty in scientific publishing; 'conventional wisdom' will be more jealously guarded than ever and only articles of 'friends' will be accepted for publication. Even now, certain journals are effectively controlled by particular groups and so only views approved of by those groups

Einstein's Biggest Blunder?

are accepted for publication; the future looks bleak indeed for scientific publishing.

The actual book by Magueijo is a peculiar mixture. It is not badly written as far as the English is concerned, apart from his too frequent descents into the language of the gutter. However, the content is rather mixed. The first part is concerned with summarizing the situation in cosmology when he first thought of investigating the consequences of a variable speed of light. An individual's opinion of summaries such as this will vary from person to person, but some of the explanations offered in this case are made unnecessarily abstruse by pointless attempts to be amusing. The essence of such a section is to transmit ideas as quickly and clearly as possible; silly attempts to attract the attention of the 'animal rights community' do nothing other than detract from the real business in hand. These days, also, when discussing the Michelson Morley experiment, which supposedly removed all possibility of the existence of a luminiferous aether, it would be sensible to note that this experiment was originally performed more than one hundred years ago. In other words, this important experiment was performed before the notion of a boundary layer was introduced formally to the world of physics by Prandtl; - the idea had been mooted by Stokes[*] many years before but its accepted introduction came in 1904 through Prandtl[†]. Why should this idea be important? Quite simply, if an aether exists, the surface of the earth would lie within the boundary layer

---

[*] G. G. Stokes, 1845, *Phil. Mag.* **XXVII**, 9
[†] L. Prandtl, 1904, Proc. 3rd. Internat. Math. Congr

between the earth and the aether. Hence, since the Michelson Morley experiment was performed on the earth's surface, it would have taken place within that boundary layer and so a null result would be expected. In other words, a valid explanation of the results obtained for that experiment would be that the existence of an aether could not be discounted by it. For the benefit of science, it would be ideal if the Michelson Morley experiment could be repeated outside this possible boundary layer. In these days of experiments being carried out in space, this is possible and would make an excellent experiment to be performed during a shuttle flight or on the international space station. However, as with most of the book, the real point of contention concerns the attitude of the writer towards those people and institutions to whom he owes his present position. Lack of gratitude and total lack of humility are the only character traits displayed both here and in the second part. If the author is as pleasant a person as some people claim, he has done himself a grave injustice through his writing.

The second part deals specifically with the work on a variable speed of light. This is jumbled and doesn't give a very good view of how things come about in scientific research generally. What is written may be an accurate account of what took place in this case, but is certainly not something which may be taken to be generally valid - as is the case with the description, early on in the book, of 'discussions' in his office in Cambridge University. Probably all generalisations of research methodology are incorrect; scientific research is for some an individual pursuit with possible collaboration with someone else, for others the essence is working in a group. Into which category a person falls is probably determined by personal temperament and/or area of interest. The main impression left by this book is of an angry, embittered young man who wants to hit out quite randomly at everyone and everything to which he has cause to be grateful. Many may have reason to criticise the people and institutions under attack, but surely this writer should be willing to express a little more gratitude to institutions such as Cambridge University and Imperial College, as well as to various individuals, for the privileged position in which he finds himself today.

The entire story so far of the variable speed of light notion is plagued by intrigue of one type or another. However, it is very much an ongoing story. Today, Magueijo and people associated with him seem to have little, or no, difficulty in publishing material concerned with theories of a variable speed of light. Others, however, still seem excluded. It is worth remembering that, in a sense, the whole question of a variable speed of light started with Einstein himself in 1911, virtually one hundred years ago. Although he seemingly abandoned the idea, it has continued to return to haunt physics, possibly because it is realised that it does offer possibilities for solving some outstanding problems and also because some of the fundamental ideas of relativity still don't rest easily with some physicists, just as they didn't all those years ago with such as Rutherford and Soddy.

Exploding A Myth

However, where does this story fit in with the notion of 'conventional wisdom'? All relativity theory is a well accepted part of modern physics. People who question its validity are regarded as being akin to heretics. When people like Thornhill question the correctness of relativity in any way, their view is one which attracts instant disapproval and is automatically regarded as being incorrect, if not actually distorted. As will be noted later, there have been examples of extremely well-known, well-respected scientists who, in later years, have raised queries about relativity and have been ostracised immediately by many who had previously been close friends and colleagues. Thornhill is not in that category. He is someone who has always harboured genuine, carefully considered doubts about the validity of the subject. Hence, the question of why his earlier thoughts on a variable speed of light did not appear in a mainline journal appears to be answered by the fact that he was opposing a major theory which enjoyed the full support of 'conventional wisdom'. Of course, this raises once again the question of how Albrecht and Magueijo managed to succeed in having their contribution accepted as it was; always bearing in mind that Moffat is said, by some, to have failed to achieve publication in a front line journal as well as Thornhill. This is undoubtedly an interesting question and one to which, no doubt, the complete answer will never emerge. It is amusing to note, however, that Magueijo, in his book[*], claims that his new theory

---

[*] J. Magueijo, 2003, *Faster than the speed of light*, (Perseus Books, U.S.A.)

Einstein's Biggest Blunder?

supports Einstein's theories, rather than the reverse, as appears to be the case at first sight. Considering the nature of his argument, many might feel, quite reasonably, that this claim needs amplification. However, evidently the doyens of 'conventional wisdom' are satisfied that this is the case. Hence, this is where, and why, this little story has a place in an examination of this hidden constraint in science, a constraint which is often termed 'conventional wisdom'.

Chapter Two

Einstein's Theories of Relativity

In the nineteenth century, the existence of a material medium, the aether, pervading all space was a generally accepted concept. The supposed mechanical vibrations of this medium were used to explain the wave propagation of light. One great challenge facing experimentalists, therefore, was to detect the actual presence of this medium. At the time, optical experiments were the most accurate available. Easily the best known was that performed by Michelson and Morley in the 1880's. It is well recorded that this experiment failed to detect the physical existence of the aether. In the history of the development of special relativity, this is the first juncture where questions should be raised. Was it actually true that the experiment did fail to detect the physical existence of an aether? The controversy surrounding this seemingly straightforward question continued throughout the twentieth century and is not resolved even today. It is claimed in the vast majority of, if not all, textbooks that no absolute motion was detected but, in truth, the published data revealed a speed of 8km/s. However, this made use of Newtonian theory to calibrate the equipment and was a figure much less than the 30km/s orbital speed of the earth. It was purely due to this second point that the detected speed was less than the orbital speed of the earth that a null result was claimed. It is now claimed by some that modern analysis leads to a different calibration for the equipment and that this, in turn, leads to a speed in excess of 300km/s. The claim is then that the experiment both detected absolute motion and the breakdown of Newtonian theory. This first supposed detection of absolute motion has supposedly been confirmed by other experiments.

However, it quickly became accepted generally that the Michelson and Morley experiment did, in fact, fail to detect the existence of an aether and there then resulted a major challenge to the theoreticians to explain this null result. After much preliminary work by such as Lorentz and Poincaré, Einstein's special theory of relativity emerged as the accepted explanation. However, since those early years of the twentieth century, there has been much discussion of the results of the Michelson-Morley experiment; it being claimed on many occasions that the experiment did not, in fact, produce a null result. The controversy still exists, to the extent that there are plans to perform the experiment yet again in an attempt to establish beyond all doubt the true facts of the situation. Nevertheless, one important piece of physics is invariably omitted from all these considerations. At the time of the original Michelson-Morley experiment and, indeed, at the emergence of the special theory of relativity, the notion of a boundary layer was unknown. Although Stokes had broached the

Exploding A Myth

idea in the middle years of the nineteenth century[*], boundary layer theory, as such, was introduced only in 1904 by Prandtl. His original publication was in an obscure journal[†] and it was quite some time before the ideas became both known and accepted.

However, if an aether did exist and if the ideas of boundary layer theory are accepted, then the Michelson-Morley experiment, since it was performed on the surface of the earth, would have been performed within the boundary layer between the earth and the aether. At the earth's surface the relative speed of earth and aether would be zero and so, on the basis of this, a null result should have been expected. Ideally, the Michelson-Morley experiment should be repeated, but this time well away from the possible boundary layer. Seemingly this would necessitate performing it well away from the earth and from all other planets. If the results of such an experiment were not null, the existence of an aether could be denied no longer and it would not be mandatory to assume the constancy of the speed of light. An important consequence would be that, as has been shown by Thornhill, the speed of light would be proportional to the square-root of the temperature of the background radiation. In turn, as has been noted elsewhere[‡], this would negate the need for the inflationary scenario in the description of the very early universe.

---

[*] G. G. Stokes, 1845, *Phil. Mag.* **XXVII**, 9
[†] L. Prandtl, 1904, Proc. 3rd. Internat. Math. Congr.
[‡] G. H. A. Cole & J. Dunning-Davies, 2001, in *Recent Advances in Relativity Theory*, vol. 2 (eds. M. C. Duffy & M. Wegener), 51

Einstein's Theories of Relativity

In a series of articles going back to at least 1985, Thornhill has revisited the whole question of the validity of the special theory of relativity. However, he has approached the question from the point of view of a fluid mechanician. Recently[*], he has concerned himself with contrasting the space-real time of Newtonian mechanics, including the aether concept, with the space-imaginary time of relativity involving no aether. By using the theory of characteristics, he showed that the usual Maxwell equations and sound waves in any uniform fluid at rest possess identical wave surfaces in space-time. Also, in the absence of charge and current, Maxwell's equations reduce to the same wave equation which governs such sound waves. This equation is not general and invariant but becomes so when transformed by Galilean transformation to any other reference frame. The same is true of Maxwell's electromagnetic equations which are not general but unique to one frame of reference; in fact, if the argument of Abraham and Becker[†] is followed through to its logical conclusion, it is seen that, in a general frame of reference, Maxwell's equations assume a form which is invariant under Galilean transformation and in which the operator $\partial/\partial t$ is replaced by Euler's total time derivative moving with the fluid, namely

$$\frac{D}{Dt} \equiv \frac{\partial}{\partial t} + \underline{u}.\nabla$$

---

[*] C. K. Thornhill, 1996, Hadronic J. Suppl. **11**, 209

[†] M. Abraham & R. Becker, 1932, *The Classical Theory of Electricity and Magnetism* (Blackie & Son Ltd., London) pp. 141-2

where $\underline{u}$ is the constant relative velocity between the two frames in question[*]. (Here by the word 'operator' is meant a mathematical symbol which indicates to the scientist a mathematical operation to be carried out on the symbols which follow.) The resulting progressive equations are then invariant and apply to electromagnetic waves in a uniform aether moving with constant velocity $\underline{u}$ relative to the frame of reference. It is what Thornhill regards as the mistake of believing Maxwell's original equations invariant which has led to the Lorentz transformation and special relativity. Also, he would contend that it has led to the misinterpretation of the differential equation for the wave cone through any point as the quadratic differential form of a Riemannian metric in space-imaginary time.

It is possibly of interest to note at this point that these modified Maxwell electromagnetic equations might be important as far as the problem of the origin of planetary magnetic fields is concerned. The mechanism generally favoured for the explanation of the origin of these fields is the so-called dynamo mechanism, although the main reason for its adoption does appear to have been the failure of the alternatives to provide a consistent explanation. Unfortunately, as far as this mechanism is concerned, in 1934, Cowling proved, in a short note in the *Monthly Notices of the Royal Astronomical Society* (vol. 94, page 39), what is essentially an anti-dynamo theorem. He showed that there is a limit to the degree of symmetry encountered

---

[*] J. Dunning-Davies, 2002, Hadronic J. **25**, 251

Einstein's Theories of Relativity

in a steady dynamo mechanism. This result, in turn, shows that the steady maintenance of a poloidal field is not possible and this has caused enormous problems over the origin of planetary magnetic fields ever since. However, Cowling's proof depends crucially on the usual Maxwell electromagnetic equations. If the above modified equations are used, the proof of the theorem does not follow and a major difficulty associated with the origin of planetary magnetic fields is removed. Of course, the major price to pay for the resolution of this difficulty is felt by most to be far too high since it involves the abandonment of much, at least, of the theory of special relativity as employed in science nowadays.

To digress slightly for a moment, it is interesting to realise that the modified form of the Maxwell electromagnetic equations referred to here has been derived independently on a number of occasions by a variety of people. Possibly most notable among these is Heinrich Hertz, whose derivation of the modified form is included in his 1893 book, *Electric Waves*[*]. This is truly notable because the date precedes relativity by so many years. Phipps[†] has queried whether Maxwell was aware of this work by Hertz and, if he was, why it didn't provoke him to re-examine his equations himself. However, it is possible, even likely, that Maxwell was aware of this work because it is known that he visited America and discussed the possibility of carrying out experiments using an interferometer to check on the

---

[*] H. Hertz, 1893, *Electric Waves*, (Macmillan, London)
[†] T.E.Phipps, 2002, Galilean Electrodynamics, **13**, 63

Exploding A Myth

possible influence of higher order terms in his theory. It is thought by some that this is what provoked Michelson to set up and perform his now famous experiment. If this speculation is true, the second part of Phipps' query remains as to why Maxwell didn't re-examine his electromagnetic equations. Of course, it is possible that he did but failed to complete a derivation in a moving medium. However, it is probably more important to note that, if Maxwell did know of Hertz's work, then others would have also and it is surprising, therefore, that special relativity came about as it did. Indeed, following Thornhill's reasoning, it may be felt surprising that special relativity, as known today, ever surfaced. The above mentioned paper by Phipps goes some way to explaining this latter query though. He points out that Hertz used a complicated component notation and didn't make use of known vector identities to simplify it. Also, he imposed an unfortunate interpretation on the velocity appearing in the expression for the Euler total time derivative which led to false predictions – for example, the prediction of the creation of a magnetic field by a moving dielectric – which were disproved soon after his death. Hence, Hertz's theory was discarded, but without a true examination of its fundamental mathematical merit. It is easy, and probably correct, to say that this was understandable but, for the future development of science, it was unfortunate to say the least. It is also interesting to note that Phipps points out that observations had been made in the latter half of the nineteenth century which raised queries relating to the familiar form of the Maxwell electromagnetic equations. Why these were ignored, but criticisms of Hertz's ideas were not, is clearly open for future speculation. In this case, however, unlike some others, both Hertz and Maxwell were internationally well-established as scientists and so, the excuse, if proffered, that Hertz (in this case) was not sufficiently well known amongst scientists of the day is simply not valid. This comment is highly relevant particularly considering the case of Waterston which also occurred during the nineteenth century.

The affair concerning Waterston and the kinetic theory of gases is well documented in Brush's excellent, and eminently readable, two volume work *The Kind of Motion We Call Heat*. Although the story has no direct relevance to the matter immediately under discussion, it is worth recalling one or two aspects of the case for reasons that will become clear later. In 1917, Schuster and Shipley[*] claimed, in their book *Britain's Heritage of Science*, that "Waterston probably furnishes the most conspicuous example of a long-continued neglect of work which would have marked a great advance in knowledge had it been recognised at the time of its maturity". In the end, it was Lord Rayleigh who eventually discovered Waterston's original article, *On the Physics of Media that are Composed of Free and Perfectly Elastic Molecules in a State of Motion*, buried in the archives of the Royal Society of London. As secretary of the said society at that time, he had little difficulty in retrieving the

---

[*] A. Schuster and A. Shipley, 1917, *Britain's Heritage of Science* (Constable, London)

manuscript and ensuring that it was published in the *Philosophical Transactions* of the society in 1892, forty-seven years after it was first submitted and, tragically, some nine years after Waterston's death. However, at this point in time, what seems particularly relevant, especially in the present context, is some of the content of Lord Rayleigh's quite lengthy introduction to the paper as printed in the *Philosophical Transactions*. He discusses the history of the paper briefly but, on page 3, states that "the history of this paper suggests that highly speculative investigations, especially by an unknown author, are best brought before the world through some other channel than a scientific society, which naturally hesitates to admit into its printed records matter of uncertain value. Perhaps one may go further and say that a young author who believes himself capable of great things would usually do well to secure the favourable recognition of the scientific world by work whose scope is limited, and whose value is easily judged, before embarking upon higher flights." This, and more in his introduction, may reasonably be viewed as a scarcely veiled condemnation of the refereeing processes in place at the time of Waterston's original submission. However, it may also be viewed as a piece of very sound advice to young researchers these days as well, - particularly if one expands his remarks to include the prestigious academic journals as well as the learned scientific societies. It does appear, however, quite clear that the cancer of 'conventional wisdom', to which reference has been made already on several occasions, has been around in learned scientific circles for quite a long time. It is, no doubt, a vain hope to think it might go away but, at least if the spectre is made public, its influence may be reduced.

In a further article[*], Thornhill showed that the equations governing general small amplitude wave motions to first order in the general unsteady flow of any general fluid also reduce to the same wave equation with constant thermodynamic wave speed in the case of a fluid at rest. The said wave equation was shown to hold in a unique frame of reference and is not, therefore, invariant under Galilean transformation. However, it emerged that it will transform under Galilean transformation into a form which is invariant for all other frames of reference. The wave surfaces of Maxwell's equations are then as for sound waves in any uniform fluid at rest. Again it follows that Maxwell's equations will hold only in a unique frame of reference and should not remain invariant when transformed into any other frame of reference. In particular, he showed that the envelope of all wave surfaces passing through any point at any time is, for the wave equation and, therefore, for Maxwell's equations also

$$c^2 dt^2 = dx^2 + dy^2 + dz^2, \qquad (1)$$

where c is the constant thermodynamic wave speed. As he pointed out, this is a differential equation and the immediate task should be to solve it; this he does. It is obvious that this equation is

---

[*] C. K. Thornhill, 1993, Proc. R. Soc. (London) **442**, 495

Exploding A Myth

$$ds^2 = c^2dt^2 - dx^2 - dy^2 - dz^2$$

with $ds = 0$. Thornhill's claim is then that this is where one mistake occurred, and has continued to occur. His contention is that there is no requirement for Maxwell's equations to remain invariant under transformation and that the above expression for ds^2 has meaning in the present context only when $ds = 0$. He suggests that Minkowski erred in apparently failing to recognise that equation (1) above is merely the differential equation of the envelope of the wave surfaces. A further point to be noted at this juncture is that Maxwell's equations, as normally considered, are derived for a medium at rest. It is conceivable that, if those equations had been derived for a moving medium originally, the controversies surrounding special relativity might never have arisen because that particular development might never have been required.

The above situation concerning Maxwell's equations and sound waves then raises the question of whether, or not, mathematics is required to tolerate the same equation being transformed in different ways for different applications. As Thornhill puts it, "does mathematics allow the wave equation to conform to Galilean transformation when it is applied to sound waves, to Lorentz transformation when it is applied to electromagnetic waves, and to either or both of these transformations when it is considered purely as a mathematical equation, or does mathematics insist that the Galilean transformation is unique and must apply equally to all equations so that the same equation must always be transformed by the same Galilean transformation, no matter to what it is applied, or whether it is applied to anything at all?"

It is recognised that the abandonment of special relativity and a return to Newtonian mechanics would result in a backlog of problems requiring conventional solutions. However, the claim is that such problems would lead eventually to the methods of unsteady gas dynamics and the theory of characteristics, such has already occurred in some instances. Thornhill himself has already tackled the problem of the kinetic theory of electromagnetic radiation and derived Planck's formula for the energy distribution in a black body radiation field from the kinetic theory of a gas with Maxwellian statistics[*]. It is in this article that he shows that, if there is an aether, the speed of light is proportional to the square root of the temperature.

In this latter paper, and in a companion one[†], he argues persuasively against another reason for denying the existence of an aether. This asserts that the Maxwell equations indicate that electromagnetic waves are transverse and so, any aether, if it exists, must behave like an elastic solid. Thornhill points out that Maxwell's equations show that the oscillating electric and magnetic fields are transverse to the direction of wave propagation and say nothing about condensational oscillations of any medium in which the waves propagate. The deduction that electromagnetic waves

---

[*] C. K. Thornhill, 1985, Speculations Sci. Tecnol. 8, 263
[†] C. K. Thornhill, 1985, Speculations Sci. Technol. **8**, 273

are transverse might be felt an alternative way of claiming the non-existence of an aether. However, if an aether does exist, then, since electric field, magnetic field and motion are mutually perpendicular for plane waves, the deduction from Maxwell's equations would be that the condensational oscillations of the aether are longitudinal, in analogy with sound waves in a fluid.

Further, as has been pointed out by Thornhill[*], the reason Lorentz 'invariance' gives so many correct results is because one consequence of the Prandtl boundary layer theory is that the viscosity of the aether ensures that the local aether moves with all observers and all observers who move with the local aether have the same unique local wave-hyperconoid given by the differential equation

$$(dx/dt)^2 + (dy/dt)^2 + (dz/dt)^2 = c^2. \quad (2)$$

This follows since the general wave-hyperconoid

$$(dx/dt - u)^2 + (dy/dt - v)^2 + (dz/dt - w)^2 = c^2$$

is invariant under Galilean transformation and, locally, $u = v = w = 0$ for all observers in their rest frames. Again, as noted already, the invariance of (2) between all observers is established by using Galilean transformation, Newtonian mechanics and the aether concept.

Hence, it would appear that there are genuine points of concern over the total validity of the special theory

---

[*] C. K. Thornhill, 2004, Hadronic J. **27**, 499

Einstein's Theories of Relativity

of relativity. However, it should not be forgotten that another major consequence of the theory was that mass and energy are related via

$$E = mc^2 \text{ and } m = m_o / \sqrt{1 - v^2/c^2}.$$

As Okun[*] has pointed out, care must be taken with the exact interpretation of these equations. Nevertheless, both have been confirmed experimentally and have proved extremely useful to practicing physicists. The theory as a whole, though, might usefully be examined afresh by open-minded people. Some concerns have been voiced here but there are others. One obvious, though rarely mentioned, concerns the region of validity of Einstein's special relativity, always assuming that that theory is accepted as correct. From the outset, it is assumed quite specifically, and Einstein himself seemed quite clear on this point as on many others, that it is the speed of light in vacuo which is assumed constant. However, how often is the claim heard that it is the speed of light which is assumed constant? This may be due to an imprecise use of language but, unfortunately, such lack of precision can lead to incorrect understanding all too easily. If the assumption is adhered to strictly, the question as to where the theory is applicable immediately becomes relevant. Unfortunately, at the very high speeds to which the relativistic corrections apply sensibly, smallish deviations from the 'in vacuo' speed will probably not

---

[*] L. B. Okun, 1987, *A Primer in Particle Physics* (Harwood Academic Publishers, Switzerland)

Exploding A Myth

prove to be significant. As in so many situations, our man-made models, although imprecise, are accurate enough to produce numerically satisfactory results. This, of course, is the reason that Newtonian mechanics works so well over such a large range of values of the speed. Even if it is correct, relativistic mechanics affects our calculations only rarely. The issue of the region of validity, however, remains very real, if only to correct misapprehensions arising in the public mind about an important area of physics. All issues, such as this, must be seen as truly important for several reasons. Uppermost here has to be the fact that much scientific research is funded out of the public purse. Surely, therefore, the public should be made aware of the truth surrounding these issues? As with so many areas of science, the pop science literature in this field, though beautifully produced and illustrated, can be frugal with the absolute truth.

After a long time spent promoting relativity, Herbert Dingle[*] raised several further worries and objections; most notably possibly that concerning the seeming non-symmetry of the problem of the so-called 'clock' or 'twin paradox'. Whatever a person's personal views may be, it is undoubtedly true that the history of this dispute (fully documented in the given reference) hardly indicates a satisfactory resolution of a genuine problem. Here, after all, was a major query being raised by one who had been a very genuine supporter of the special theory of relativity as put forward by Einstein

---

[*] H. Dingle, 1972, *Science at the Crossroads*, (Martin Brian & O'Keeffe, London)

Einstein's Theories of Relativity

and, once again, a person well-known and well-established in academic circles. Dingle experienced real concerns over the validity of the theory and, as well as those, he recognised that there were in existence *two* special theories of relativity, one attributable to Lorentz and the other to Einstein. The difference between the two, as he pointed out, was a big one; the first retained the concept of an aether, the second did not. Possibly the most worrying aspect of the case of Dingle is the attitude of fellow scientists to his persistent querying. All recognise that, if someone continues returning to the same old question regardless of the reply given already in hopeful answer, patience can wear a little thin. However, in this respect, one is reminded of the awkward questions little children so frequently ask. Such questions are seemingly trivial - indeed that is how they sound superficially – but they often prove very difficult, if not impossible, to answer and not only because the explanation is too difficult and complicated for a small child to understand, but because the apparently 'all-knowing, all-understanding' adult simply doesn't know the answer. This is a situation familiar to all who have been privileged to be close to young children. All one can sensibly do is to tell the absolute truth. Admitting that you don't know, can be felt to be degrading, even a little humiliating, but, on occasions, it is the only course of action which can be followed honourably. When one has read and digested what is included in Dingle's book, *Science at the Crossroads*, the conclusion has to be reached that a full, frank and totally open discussion of the points raised would have been in everyone's interests. The attitude,

Exploding A Myth

so prevalent in society today as then, of someone adopting the stance of stating that they have given their answer and, if you have the temerity to contact them again, they will simply file your communication but not reply, is totally unacceptable in any area of normal society. Hence, how much more unacceptable must it be in scientific circles, where all concerned are supposed to be seeking scientific truth? However, this certainly appears to have been the attitude faced by Dingle. No-one can say what the outcome of an open, detailed discussion would have been but it is certainly true to say that the air would have been cleared and all the problems would have been in the public domain. Such an outcome could not have achieved anything other than good for science as a whole. Members of the general public may not understand the finer points of detailed abstruse science, but they understand and appreciate far more than is, on occasions, thought. Too often, patronising people by assuming they will not understand something leads to problems. There have been numerous cases, particularly in the field of public health, where imprecise information has been released and distrust of the profession as a whole has resulted. Another result of such an approach can be the rise in influence of 'quacks', for want of another word. In all areas of science, frank full open discussion is the only real way to make any sort of genuine progress. As far as special relativity is concerned, it has to be said that the jury is still out. The present situation is, in the view of many, a very unsatisfactory one for science but will not be resolved until all problems are laid out in public and discussed fully and openly in the public arena. This will require an enormous 'volte-face' by a great many people.

However, returning to Einstein and his theories once again, it should never be forgotten that he also thought very deeply about the problem of gravitation. Whether or not he turned his attention to this because of the problem of the unexplained shift in the perihelion of the orbit of the planet Mercury is not really important, although it does provide a convenient starting point for any discussion of what is now known as Einstein's General Theory of Relativity. The name merely indicates a follow-on from his special theory but, in fact, it is really a theory of gravitation although, like all others theories of gravitation, it doesn't explain exactly what the force of gravity really is. The final point is not at all surprising since, as mentioned elsewhere, no-one really understands what a force is, merely what it does! It is often pointed out that people such as Poincaré and Lorentz contributed greatly to the special theory of relativity but, where the general theory is concerned, the tremendous intellectual achievement was Einstein's own. True he made use of the mathematical results of such as Riemann, Bianchi and Ricci, but the final physical theory was entirely the work of Einstein himself; he merely made use of known results in differential geometry in the same way as others utilised known results in algebra or calculus. As well as explaining the problem associated with the orbit of Mercury, the theory also made predictions concerning the bending of light rays as they passed a massive body such as the sun. This offered almost immediate scope

Exploding A Myth

for scientists to test this revolutionary new theory. The eclipse of 1919 provided the perfect opportunity. The observations made of this eclipse by Eddington were used to herald the almost complete vindication of this theory, although subsequently doubts have been cast over the actual information obtained at that time. Dingle[*] points out that, up to the time that Einstein's general theory was brought to everyone's attention as a result of the eclipse observations, to most people, the theory of relativity meant Lorentz's theory. It was only after the events of 1919 that Einstein's theory of special relativity gained prominence. As has been pointed out by Dingle also, both Lorentz and Einstein knew the difference between the two theories, but few others did or, for that matter, do now. In the early years of the last century, people of the academic eminence of Ritz, Lodge and Poincaré all seemed to regard Lorentz as the originator of what they understood the principle of relativity to be, and, as mentioned earlier, Lorentz's relativity retained an aether. Hence, it would appear that, historically, the acceptance of Einstein's general theory had a profound effect on the entire scientific community as far as the special theory of relativity was concerned. Also, since that time it seems that, to even mention the word 'aether' is to court scientific banishment; people who use it seem to be regarded as the pariah's of true scientific society; but why? It is merely a word. Why should it be excluded from our vocabulary? This may appear a trivial point and, indeed, it may be but, throughout present day scientific

---

[*] Ibid

literature, other words are used quite regularly which might easily and sensibly be replaced by this ostracised word 'aether'. The most popular alternative is the word 'vacuum'. This has now come to the fore and it does not mean what it used to mean or, at least, if it does, that original idea has been expanded and modified out of all recognition. The vacuum is now regarded as possessing a great deal of structure, to the extent that the book *The Structured Vacuum* even lists seven seemingly quite distinct special vacua, apart from the usual vacuum. The dielectric vacuum, the charged vacuum, the opaque vacuum, the melted vacuum, the grand vacuum, the Higgs' vacuum, and the heavy vacuum are all listed separately and granted separate chapters in which to be discussed. Of course, these seven are all seen to refer to different aspects of the actual vacuum but, seeing this list of, at first sight, seven different vacua, brings home quite forcibly the realisation that nowadays the simple idea of the vacuum has progressed to unimaginable new heights of complexity and, possibly even, abstraction. However, this vacuum is, in a sense, an all-pervading medium and so, in some ways at least corresponds to the original ideas surrounding the notion of the aether.

A further point which might be noted concerning modern ideas of the aether, as distinct from the vacuum, is that the actual proposed mass of an aether particle is claimed by Thornhill, amongst others, to be of the order of 10^{-39}kg. This is a figure which is being heard more and more often in physics circles. Is it, in fact, a figure for a mass which really does have a genuine far reaching significance? At present, no-one knows the

Exploding A Myth

answer to that question but it is a real question which, hopefully, will eventually produce a real answer.

At this point, it would seem sensible to consider the work of Harold Aspden. Given Einstein's background, it is not a little ironic to note that, in a brief biography on his website (www.aspden.org), Dr.Aspden reveals that he is a retired corporate *patent* director who has a special interest in physics. He notes that it was in 1969 that he published a very interesting little book entitled *Physics without Einstein*[*] and comments that it received little real attention from the general physics community, no doubt because of its title. He has now completed a more up-to-date text entitled *Physics Unified*, which is available on the above-mentioned website. This again aims to explore the revolutionary idea that physics could have progressed further and more fruitfully without Einstein. This indeed seems to have been the theme underlying most, if not all, of Dr. Aspden's work in physics – an unswerving belief in the damaging effect of Einstein's basic scientific philosophy on the progress of physics in the twentieth century. Such a belief will be regarded as akin to blasphemy in many circles but can his views be dismissed so easily, can they be simply ignored? It is certainly worth reviewing his views and suggestions but, more particularly, the basis for them.

Over the years, Aspden has produced so much interesting and relevant material that it is difficult to

---

[*] H. Aspden, 1969, *Physics without Einstein*, (Sabberton Publications, Southampton)

Einstein's Theories of Relativity

know where to start; what should be included, what excluded? However, early in his later writing[*], he reveals some very interesting facts which, while probably well-known to some, will, I suspect, be far less well-known to the vast majority. He points out that physics, particularly electrodynamics, made tremendous and very rapid progress in the later years of the nineteenth century. One of the highpoints of this had to be the discovery of the electron by J.J.Thomson in 1897. This, of course, is well-known but what is less well-known is that this was followed, in 1901, by Kaufmann's discovery[†] that the electron's mass increased with speed. In fact, Kaufmann actually measured variation in the charge/mass ratio with increase in speed. The immediately obvious point concerning this piece of information is that it clearly predates Einstein's 1905 paper introducing his special relativity. It is also worth noting, because it is often either forgotten or deliberately ignored, that the explanation for this variation with speed had been provided by Thomson and others before the advent of Einstein's special relativity. Aspden has obviously delved very deeply into the scientific history of the now famous formula linking energy and mass and this is to the benefit of all, whether or not individuals agree with his conclusions. He notes that, as far as the formula $E = mc^2$ is concerned, definite reference was implied in a book of 1904, - *The Recent Development of Physical Science* by W.C.D.Whetham - where there was

---

[*] H. Aspden, 2005, Physics without Einstein; A Centenary Review, (see www.aspden.org)

[†] Kaufmann, 1901, Gottingen Nach. **2**, 143

reference also to a suggestion made by Jeans to the effect that the energy of radioactive atoms might be "supplied by the actual destruction of matter". In other words, in an article of 1904 published in *Nature* (vol.**70**, page 101), Jeans directed everyone's attention to the store of energy which was available by the annihilation of matter, "by positively and negatively charged protons and electrons falling into and annihilating one another, thus setting free the whole of their intrinsic radiation". Jeans further noted that, initially, he felt he was advocating something new but actually found that Newton had anticipated something similar two centuries earlier, as is recorded in Query 30 of the 1704 edition of *Optics*. However, returning to the question of the equation $E = mc^2$, as Aspden notes, while specific reference to it does not appear in Whetham's book, all the necessary background physics is well presented in mathematical terms. No doubt, Thomson had arrived at his result by assuming the energy of the magnetic field due to the motion of a charge e at a speed v to be $e^2v^2/3ac^2$ and thinking of this as equalling the kinetic energy $mv^2/2$. The equality of these two expressions results in:

$$mc^2 = 2e^2/3a,$$

where the expression on the right-hand side is the energy Thomson recognised as that of an electron with its charge contained within a sphere of radius a. Hence the implied equivalence of mass and energy is deduced. Once again, Rayleigh's comment relating to Waterston comes to mind. Just as in the earlier case of Hertz, the personnel being considered here were truly eminent

Einstein's Theories of Relativity

men of science and were so at the time they made their separate suggestions and yet these suggestions obviously did not receive due recognition at the time they were made and have been largely – almost totally as far as most are concerned – ignored since.

In Britain, this sort of behaviour almost seems to be a national malaise. It is not uncommon for non-British achievements to be given almost excessive acclaim and publicity, while the earlier similar British achievement is consigned to obscurity. A perfect example of this is the discussion in chapter one of the uncompromising attitude of ignoring Thornhill's work relating to a variable speed of light. It is of further interest to note that, in all these cases, the earlier work has been published and, as far as Jeans is concerned, since he published in *Nature*, it is difficult to imagine the reason for the neglect of his work being that he had published in an obscure journal; it might be noted further that this remark carries even deeper significance when the date of Jeans' publication is considered – in those days, the present plethora of academic journals simply did not exist!

However, be that as it may, there is one further important point raised by Aspden in the context of the validity of Einstein's special relativity and it is concerned with the concept of an aether. Again Aspden has successfully and usefully delved into history and revealed another set of facts which are largely unknown today; certainly no undergraduate will find this information in any conventional recommended textbook. He notes that the nineteenth century physicists faced a dilemma in that no-one knew whether the aether, though invisible, was a fluid or solid medium. This problem had been complicated by the existence of a theorem of 1839, due to Samuel Earnshaw. To quote Aspden, it seemed that Earnshaw "recognised the need for the aether, if composed of electric charges, positive and negative, in equal numbers, to define a kind of crystal structure as a frame of reference for light propagation", However, he 'proved' that "such a stable structure was impossible given our understanding of the law of force known to be operative between discrete electrical charge forms. If the aether existed, chaos had to prevail, there being no orderly form that could be possible". Earnshaw's theorem that

A charged body placed in an electric field of force cannot rest in stable equilibrium under the influence of the electric forces alone,

appears, with proof, on page 167 of Jeans book, *The Mathematical Theory of Electricity and Magnetism.* However, as is shown by Aspden[*], Earnshaw had made the mistake of assuming the aether composed solely of electrically charged particles sitting in a void. It is possible to think of an aether composed of a continuum of uniform electric charge density but containing discrete charges of a total charge sufficient to render the aether electrically neutral overall. While adherence to this theorem had grave effects for the aether concept, it also had a negative influence on the advance of

---

[*] Ibid

elementary particle theory. On page 168 of the above-mentioned book, Jeans used the Earnshaw theorem to place strict limitations on the structure of matter; limitations which would be reasonable if, but only if, the Earnshaw theorem was valid. As Aspden points out[*] the work of Earnshaw would seem to rule out the modern notion of particles, such as protons and neutrons, being composed of quarks. Why has it taken so long for the flaw in Earnshaw's theorem to be highlighted? This is an obvious and relevant question to which no truly honest answer may be given. It is a theorem which, as far as many are concerned, doesn't even exist, but its effects on the advancement of physics have been far-reaching and not to the benefit of that discipline. However, although the error in the proof has been pointed out now by Aspden, it should be noted that it was certainly known to W. T. Scott, who wrote about Earnshaw and his theorem in the *American Journal of Physics* (volume **27**, 1959, page 418) and in a book, *The Physics of Electricity and Magnetism*, published by Wiley in 1966. Hence, it seems possible, even likely, that one hundred and twenty years elapsed before anyone queried the validity of Earnshaw's theorem. To date, the answer to the question 'Why?' remains a mystery.

Since the heady days of the 1920's, relativity – both the special and general theories – has appeared to dominate physics. As far as the special theory is concerned, it is undoubtedly true that controversy has simmered just beneath the surface from the very early

---

[*] Ibid

Exploding A Myth

days. The general theory, however, seemed to offer the only solution to problems which had been taxing theoreticians for some time. Doubts were expressed but, as has so often been the case where Einstein's theories of relativity are concerned, the doubters were regularly dismissed as mere cranks. Again, though, as in the case of special relativity, not all the facts are made readily available to modern day audiences. In Newtonian mechanics, although not specifically mentioned usually, the effects of gravity are assumed to propagate at infinite speed. This follows from Newton's original concept of action-at-a-distance. More recently, the thought has developed that, in reality, gravitation propagates at the speed of light. The example that originally caused problems was, again as mentioned earlier, the value of the observed advance of the perihelion of the planet Mercury. Newton's theory explains an advance of the perihelion but not of the observed magnitude. It is proclaimed nowadays that Einstein's general theory of relativity was the first to explain the advance correctly. It is true that it does predict the correct value for the advance but, as Aspden reveals[*], Einstein wasn't the first to offer a satisfactory explanation. This honour falls to a German schoolteacher, Paul Gerber, who presented a theoretical argument giving the precise value of the anomalous advance of the perihelion of Mercury in an article entitled *The Space and Time Propagation of Gravitation* and published in 1898[†]. Gerber actually

---

[*] Ibid
[†] P. Gerber, 1898, Zeitschrift f Math, u Phys., **43**, 93

derived exactly the same formula for the advance as that given by Einstein in 1916 and, in fact, had assumed that the effects of gravity propagated with the speed of light, in common with ideas of today. Aspden comments at this point that Gerber may have made mistakes in his argument but implies that the basic argument was correct and all that was needed was for someone to tidy it up. Instead, this work was, and still is, virtually unknown. This is surprising because the article addressed a major problem of the time and the fact that it appeared in German would have posed less of a problem to international audiences then than it might now.

The arguments surrounding the advance of the perihelion of Mercury and other phenomena supposedly explained by the general theory of relativity and only by that theory have continued apace ever since the theory first saw the light of day. Most suggested alternative explanations have been dismissed, often with a sad shake of the head as if to suggest some degree of sympathy for someone who could be so deluded as to think they could even contemplate offering an alternative. Nevertheless, in more recent years, alternative ways of explaining the shift of the perihelion of Mercury and the bending of light rays have emerged. One of the most recent is that due to Lavenda[*]. He set out to explain the time delay in radar echoes from planets, the bending of light rays, and the shift of the perihelion of Mercury via Fermat's principle and the phase of Bessel functions. It is undoubtedly true that he

---

[*] B. H. Lavenda, 2005, Journal of Applied Sciences, **5**(2), 299

Exploding A Myth

has succeeded in explaining these three phenomena by this means. However, he has met fierce opposition when it comes to publishing this work. Why? Nowhere does he claim to be attempting to usurp the position of general relativity; he merely wishes to point out that some results, at least, may be obtained by means other than use of the general theory of relativity. As he himself says, "Sometimes new insight can be gained by looking at old results from a new perspective." This highly perceptive suggestion by Lavenda might usefully be noted by all who oppose the publication of anything that even appears to question either special or general relativity, or indeed any who oppose publication of anything purely because it fails to conform to some dictat imposed by some arbitrary component of 'conventional wisdom'. The alternative suggests an amazingly blinkered view, often by some of the publicly acknowledged giants of the scientific world. Whether these people are really scientific giants is immediately brought into question. The only way forward in any pursuit of knowledge is to admit all possibilities. Once you close one door, you instantaneously rule out one avenue of approach and, therefore, possible advance. Intellectual giant though Newton undoubtedly was, everyone is quite happy to query details of his theories, and rightly so. Hence, why is questioning of Einstein's theories regarded by so many in influential positions as totally unacceptable? From what one reads of the man, that is not a reflection of the position he might have been expected to espouse. Also, it is interesting to note that the same attitude does not seem to affect Newtonian mechanics. Of course, Newtonian mechanics is now extremely well-established and is the theory which dominates everything mechanical seen by the majority of people. It is eminently successful. However, no-one seems to have been offended by the analytical approach to the subject as advocated by Lagrange and Hamilton; no-one seems to have been offended by the 'forceless' mechanics suggested by Hertz as expounded in his book *The Principles of Mechanics*[*]. Why then are so many so apparently over-protective of Einstein's theories of relativity? This is a question to which no-one probably knows the true answer. Nevertheless it is a question which needs to be raised and one of which the public at large should be aware. To emphasise a point raised above, alternative approaches do exist which lead to the solution of problems which may also be solved using the methods of general relativity and, as Lavenda has said, examining these alternatives could lead to new insights. It seems reasonable, therefore, not to simply dismiss these with a sad nod of the head.

For mathematicians, the general theory of relativity is regarded as a thing of real beauty. This is a position which any non-mathematician may find extremely difficult to comprehend but it is, nevertheless, very true. It must always be remembered that mathematics is a subject which may be studied on at least two very different, but equally important, levels. It may be studied as a purely academic subject in its own right. In this approach, the mathematics is all important and, to

---

[*] H. Hertz, 1956, *The Principles of Mechanics*, (Dover, New York)

the practitioner, can be, and often is, extremely beautiful. It must be noted also that, academically, this approach to mathematics is fully justified; it is a highly worthwhile academic pursuit. However, the second major view of mathematics is as the language of physics. In this context, mathematics may still be seen as extremely beautiful but here it is, and indeed must be, subservient to the physics in importance. Once mathematics is used as the language of physics, it is being used as a tool in an attempt to describe physical situations. It is no longer truly important in its own right. Now, it is the physics of the situation under consideration which is all important and must provide the driving force for any work which ensues. Again, the mathematics is being used in this case to help model a physical situation and it must be remembered always that that is all that is being attempted – to produce a model of a physical situation. It is highly unlikely that any such model will be an exact representation of physical reality; it will be merely an approximation. How good that approximation proves to be is determined by what follows from the theory. Does it, for example, make valid predictions about the physical situation which originally occasioned the investigation? If it does, the accuracy of these predictions will prove a useful guide to the worth of the theory. However, where great care must be taken is in ensuring that the physical situation under consideration isn't, in any way, forced to 'fit' this theory; it is vital to avoid the accusation that observations are interpreted with the predictions of the theory in mind. This sounds a very trivial point to be making but nothing could be further from the truth. In these days when so much money is invested in some areas of research, there must be tremendous pressure on investigators to produce 'proof' to support the ideas put forward, probably very forcibly, in the original research proposal which led to the provision of money which, in turn, allowed the project to exist. Again, it should be remembered that, nowadays again, there is great pressure on all researchers to gain research grants for a wide variety of reasons: such grants can enhance a person's personal reputation; they can increase the profile of that person's own department within its home institution; they can increase the research profile of the person's department both nationally and internationally, which is so important in these days of research assessment exercises. All these are extremely heavy extra weights placed on the shoulders of researchers – especially the young researchers just starting out on their careers – and *not one* is beneficial for real, worthwhile research! No doubt, this pointless extra pressure will produce some good work, but that will simply be the exception which proves the rule. People pursuing topics out of pure interest will, in all probability, produce far more work of lasting value to mankind. Pressures will always be there to create a sense of urgency when it is truly required, such as in the search for cures to various medical problems, but, generally, keeping such pressure to a minimum will prove beneficial in the long run.

The general theory of relativity is one of those topics which rely heavily on very beautiful mathematics, to the extent that the physics of the situation can even tend to

Exploding A Myth

be obscured by that very mathematical beauty. Mathematics is a beautiful, rewarding subject in its own right and, academically, no justification is needed to support its study. However, as mentioned above, where study of physics is concerned, mathematics is simply a tool to be used by the physicist in aiding the resolution of a physical problem. In these circumstances, it is the physics which is all important. A theory cannot be adopted to the exclusion of all others simply because the mathematics is so beautiful. As far as general relativity is concerned, as has been stated on several occasions, the only results which can be truly trusted are those with a Newtonian analogy. It must be remembered also that, in practice, the results of the theory are used only rarely where descriptions of the physical world are involved; the results are used far more frequently to speculate about the physical world, especially its origins. One must wonder about the worth of speculating about the physical world and its origins on the basis of a purely abstract mathematical theory – however beautiful the mathematics may be. Some of these speculations, which dominate much present day thinking, involve the imposition of a physical meaning to a mathematical singularity. Both the notions of the 'Big Bang' and of relativistic black holes fall into this category. These two major issues will be addressed in the following sections.

Chapter Three

Big Bang Theory - Controversial or Not?

In an earlier chapter, attention was drawn to the fact that one supposed reason for Albrecht and Magueijo considering the abandonment of the icon of a constant speed of light was to assist with the resolution of some problems associated with the Big Bang theory for the beginning of the Universe. However, one may, not unreasonably, enquire "Why not look at the case for the validity of the Big Bang as the true explanation of the beginning of the Universe?" Of course, to do this would be to place another of the icons of modern physics under the microscope. In these days when 'conventional wisdom', rather than physics, seems to have the greater influence, to raise such a question might be too much to expect. However, if one really examines how much we *truly* understand about our world and all that exists in it and if we are completely honest, one is forced to admit that, as far as *real* understanding is concerned, our somewhat puny human minds have achieved very little, - even allowing for the towering intellects of such as Newton! True understanding of basics still eludes us. For example,

Exploding A Myth

what do we mean by the force of gravity? We appreciate and can describe the *effects* of that force but do we really know precisely what it is and what causes it? No! The same is true of all other forces also. Indeed, at the deepest level, do we really understand what any force actually is? We struggle to understand the Universe around us but find ourselves hampered not only by our own ability, or lack of, but also by the constraints of 'conventional wisdom', which effectively prevents the asking of so many questions.

The Big Bang as a valid model of the Universe has been under close scrutiny almost since it was proposed and many of the queries concerning it remain. These queries tend to be 'swept under the carpet' but in a rather subtle way. The rise of popular science books has provided a means whereby the general public is persuaded to believe in the ideas accepted as founding 'conventional wisdom'. The 'solutions' to various problems are presented as indisputable facts; the notion that other possible explanations exist is carefully suppressed. One notable exception to this observation, although it probably wouldn't be considered a 'popular' science book, is the *Cambridge Encyclopædia of Astronomy*, which appeared in 1977. This book provides an excellent example of a book which, while apparently supporting the commonly accepted view of things, nevertheless obviously leaves the door open for other explanations of observed phenomena. In many ways, it provides an object lesson in open-minded, unbiased writing of a scientific text - popular or otherwise.

Big Bang Theory – Controversial or Not?

The whole idea of the Big Bang goes back to the theoretical investigations of Alexander Friedmann[*] and Georges Lemaître[†] in the earlier years of the last century following Einstein's publication of his General Theory of Relativity. Its movement to a position of prominence, if not pre-eminence, in cosmology might be felt to have been brought about by its eloquent advocacy at the hands of George Gamow[‡] in the mid to late 1940's, ably supported by such as J. Robert Oppenheimer. It is quite widely claimed that the standard big bang model makes three major predictions which have been verified observationally. If that were true beyond all reasonable doubt, it would indeed be a theory to take very seriously. However, are these claims unquestionably true? First, it is claimed that the model predicts distant galaxies receding from one another at speeds proportional to the distance between them. This view is supposedly supported overwhelmingly by Hubble's discovery of the redshift of light from celestial objects in the 1920's. Secondly, the model is claimed to predict the existence of background radiation which is seen as a remnant of the original big bang. Support for this comes from the detection of the cosmic background radiation by Arno Penzias and Robert Wilson in 1965[§]. Some also claim that the recent examination of the properties of this background radiation by the COBE satellite again confirm totally the predictions of the big bang. Thirdly, the model is

---

[*] A. Friedman, 1922, Z. Phys., 10, 377 1924, Z. Phys., 21, 326
[†] G.Lemaître, 1927, Ann. De la Societe Scientifique de Bruxelles, 47, 49
[‡] G.Gamow, 1946, Phys. Rev., 70, 572
[§] A. Penzias & R. Wilson, 1965, Ap. J., 142, 419

Exploding A Myth

said to predict successfully the abundances of the light elements such as helium, deuterium and lithium. At the same time, these claims are taken to imply that no other theory can explain these phenomena and there are no doubts about these deductions from the basic idea of the big bang. It goes almost without saying that the interpretation of experimental and observational results which leads to confirmation of the 'truth' of the big bang theory is accepted without question. However, is the situation quite as clear cut as that? Are all the questions answered, and answered both successfully and correctly?

Now return to the beginning of the story of the development of modern cosmological thought in an attempt to understand how the present position has evolved. As far as modern ideas are concerned, one of the first major advances came with Hubble's evidence that three nebulae, M31, M33 and NGC6822, were to be found at distances far beyond the remotest parts of our own galaxy. It was accepted that these were totally separate from the Milky Way. Not long after establishing that these nebulae were extragalactic systems, he also showed that the redshift of their spectral lines increased with distance. Utilising the most obvious interpretation of redshift, that is that it is a Doppler shift occasioned by the recession of the source, it is easily seen that Hubble's result may be taken to indicate that the Universe is expanding and the most distant galaxies are receding fastest. By looking at things in reverse, this is seen to mean that the Universe was much denser in the past and there is a tendency to extrapolate back to claim that, at some distant time, all the matter in the Universe was so highly compressed that it was all confined to a single point! It is at this point in the discussion that the '*Cambridge Encyclopædia of Astronomy*' comes into its own as far as fair, scientific examination of this is concerned. It claims that, at this point, care should be taken, since, "it is possible that the simple interpretation of the redshift is not correct, and that the expansion is illusory." Even if the fact of expansion is accepted, "it does not necessarily follow that the Universe was denser in the past than now, for implicit in that conclusion is the assumption that matter in the Universe is neither created nor destroyed." However, it is pointed out also that the hypothesis that the redshift is a Doppler shift occasioned by recession of the galaxies is acceptable scientifically since it is consistent with the known laws of physics. Reflecting the time of writing, it is claimed that "no other scientifically acceptable hypothesis has yet been proposed" but it does note that, as far as the position existing at that time was concerned, there was no proof that that was the true explanation.

The encyclopædia article continues by noting that, since the time of Hubble's original hypothesis, many more observations had been made which served to confirm his postulated relationship between distance and velocity of recession. It is claimed that no obvious deviations from the simple linear relationship,

$$\text{Velocity} = \text{Hubble parameter} \times \text{distance in megaparsecs},$$

Exploding A Myth

have been detected. However, this is a point to which attention must return.

Hubble also spent a considerable amount of time investigating the distribution of galaxies in the Universe. Obviously, such observations were restricted by the instrumentation available but, nevertheless, he noted that, on very large scales, the Universe does appear homogeneous; there is no obvious sign of diminution of numbers of galaxies as the accessible limits of the Universe are approached. Also, the Universe was found to look more or less the same in all directions and the cosmic expansion seemed to be proceeding at the same rate in all directions; that is, the Universe is said to be isotropic. All this is taken to mean that there is no meaningful centre for our Universe and as confirmation that our own galaxy, the Milky Way, certainly occupies no privileged position within the Universe. Strong confirmation for the isotropic nature of the Universe is felt to be provided by the so-called cosmic background radiation, a component of radiation found by radio astronomers which is itself isotropic to a very high degree and is inexplicable as noise within receiving systems or as originating from any known radio sources. This radiation is, of course, that background radiation mentioned earlier. Since Hubble's time, however, observing equipment has changed for the better and systems are now observed quite regularly which emit far more radiation than many of those observed by Hubble. One important class of objects to be considered here is provided by the quasars; the most 'distant' quasars are thought to have

Big Bang Theory – Controversial or Not?

redshifts far in excess of those for the furthest galaxies. It is accepted by many that there were far more quasars and, indeed, radio galaxies in the past than there are now. This, if true, implies that, in the past, the Universe was different from now and this seems to pose a serious problem for the Steady State Theory of the Universe, as well as offering extremely strong support for alternatives, especially the Big Bang. However, this whole question is, or should be, a completely open one. Many seem to give the impression that everything in this area is absolutely clear cut and anyone opposing the generally accepted view is to be ignored as lacking in understanding of the truth. Frankly this appears to be the view adopted in the corridors of conventional wisdom towards the work and ideas of Halton Arp. From all that one hears and reads, it seems that Dr. Arp was, until relatively few years ago, regarded as probably one of the foremost professional astronomers in the world. He had been awarded the Helen B. Warner prize, the Newcomb Cleveland award and the Alexander von Humbolt Senior Scientist Award. He had worked at such prestigious establishments as the Mount Palomar and Mount Wilson observatories and, whilst there, had produced his catalogue of 'peculiar galaxies', by which are meant galaxies that do not possess the standard, symmetrical form of most galaxies. However, while able to make use of the most powerful of telescopes, Arp also discovered that many pairs of quasars, or more correctly quasi-stellar objects, which possess extremely high redshift values appear to be associated physically with galaxies having much lower redshift values; galaxies, in fact, which are

Exploding A Myth

known to be much closer to the earth than the redshift values of the quasars concerned would imply. This all follows from the Hubble law which indicates that objects having high redshift values must be receding from the earth very quickly and, therefore, must be found at large distances from the earth. Hence, Arp was faced with the intriguing question of how objects with totally different redshift values, objects which according to 'conventional wisdom' had to be located at totally different distances from the earth, could be physically associated – in some instances, Arp's photographs seemed to show a physical bridge between the quasars and what, to him, appeared to be the associated galaxy. As has been recorded many times, Arp has many photographs of pairs of quasars, with high redshifts, symmetrically located on either side of low redshift galaxies. It has to be noted that these pairings occur far more often then the probability of random placement would allow. Of course, the main problem with Arp's photographs is that according to Big Bang theorists, high redshift objects must be at a great distance from the earth; to them high redshift is effectively a measurement of distance from the earth. It is often claimed by the advocates of 'conventional wisdom' that Arp's statistical analysis is in error; after many years, this still seems to be the main line of attack on his work. Occasionally, this is modified to a claim that the statistical basis for his results has never been presented clearly. However, it has been said that, as far as his original article on this intriguing problem is concerned, it was sent by the journal editor to two referees – one an eminent astronomer, the other an eminent statistician. The astronomy referee supposedly reported the astronomy to be impeccable but claimed he could not comment on the statistics involved because he wasn't enough of an expert in that field. On the other hand, the statistics referee is supposed to have reported that the statistics was impeccable but he could not comment on the astronomy because that wasn't his area of expertise. On the basis of these two reports, the editor published the article. Is this story true or apocryphal? Only those immediately concerned know the answer to that question, but it certainly seems true that there is no satisfactory foundation for criticising Arp's work on the basis of the statistics involved, and that seems to be the only criticism actually offered.

Much of Arp's work is well-documented in his book *Seeing Red*. In that book he both lists and discusses many examples in support of his basic thesis and that thesis raises fundamental questions about the commonly accepted interpretation of the redshifts of astronomical objects. He is able to produce numerous photographs of galaxies with symmetrically placed quasars of very much higher redshift values seemingly physically linked with galaxies of lower redshift values. The photographic evidence appears extremely powerful and it is easy to see that there are questions which need to be both posed and answered. The answer advanced by Arp and people associated with him is that the observed redshift value of an object is composed of two parts – the usual velocity component but an additional intrinsic component also. Conventional astronomy only recognises the velocity component. The intrinsic component is definitely not associated with the body's velocity but might be thought the name for any other contribution to the total redshift of an object; a contribution which is mistaken as being due to the body's velocity and hence leads to an overestimation of that body's distance from earth. This contribution even appears to change with time, possibly in discrete steps. If the two possible components of the redshift are denoted by z_v for the normal velocity-dependent redshift and z_i for the intrinsic component, the two are added together according to

$$1 + z = (1 + z_v)(1 + z_i)$$

where z represents the total redshift. This would reduce to

$$z = z_v + z_i$$

for small redshift values.

The photographic evidence to support the assertion that quasars are, in fact, linked to parent galaxies via physical 'bridges' is compelling and may be explained by supposing the quasars to have been ejected violently from the 'parent' galaxy. From the perspective of the earth-bound observer, in this scenario, the quasar would have been ejected at a large velocity and an enormous extra component would have been added to its normal redshift value. The immediate response to this would obviously be that, statistically, it might be expected that roughly half of the quasars should be ejected towards us and half away. However, Arp frequently discusses pairs of quasars placed symmetrically relative to the 'parent' quasar. Each member of such a pair will possess the same intrinsic redshift and should have velocity redshifts of equal magnitude and opposite direction. If the two total redshifts of the components of such a pair are z_1 and z_2 respectively, the above relation leads to

$$1 + z_1 = (1 + z_i)(1 + z_v)$$

and

$$1 + z_2 = (1 + z_i)(1 - z_v).$$

A little straightforward algebra then shows that

$$z_i = (z_1 + z_2)/2.$$

Hence, the intrinsic redshift of a pair of symmetrically placed quasars is merely the arithmetic mean of the individual measured redshift values. Obviously, this simple manipulation assumes that the individual quasars possess velocity dependent contributions of equal magnitudes but opposite directions but it does offer a consistent picture of what might be happening. This is best illustrated by drawing on an example from Arp's own book, *Seeing Red*. On page 15, he cites data associated with the central galaxy NGC4258. For the pair of quasars concerned, the total redshifts are 0.40 and 0.65. If it is assumed that both have identical intrinsic redshift components, the above relation leads to

$$z_i = (0.40 + 0.65)/2 = 0.525$$

The values of the associated velocity redshifts are then, for the quasar with intrinsic redshift 0.40

$$1 + z_v = 1.40/1.525 = 0.918 \Rightarrow z_v = -0.082$$

and for the other quasar

$$1 + z_v = 1.65/1.525 = 1.082 \Rightarrow z_v = 0.082$$

Hence, as expected really, the theory does lead to a situation where the one quasar is actually approaching the earth, while the other is receding from it.

As indicated, this work of Arp's has not been welcomed by the orthodox astronomical community with open arms. Why not? Basically Arp's work, if accepted, casts severe doubt on the assumption, which is quite basic to Big Bang theory and, therefore, to most if not all of accepted cosmological theory that objects possessing a high redshift must be far away from the earth. Hence, all the claims of the Big Bang model which depend on the orthodox interpretation of redshifts must be examined afresh. Again, Arp's hypothesis, backed by such eminent physicists as Hoyle, Burbidge and Narlikar, casts doubt also on the notion that black holes lurk at the centre of quasars. As will be discussed shortly, no black hole has yet been identified beyond reasonable doubt, but, if one did exist, it is assumed that it would be drawing matter to itself rather than ejecting it at very high velocities. So once again, Arp displeases the establishment by proposing a solution to a very real problem which suggests matter being ejected from a central mass rather than absorbed into it. In much current astronomical literature, there seems to be a preoccupation with the death of stars and, in some ways more importantly, with the colliding or merging of galaxies. Arp's view, and one supported by Hoyle and many of his associates, is

Big Bang Theory – Controversial or Not?

that it is rather the birth of galaxies that is being witnessed; instead of viewing and contemplating possible collisions, it is rather separations that are being seen. It might be felt that this view is more in keeping with Big Bang cosmology in that the Big Bang supporters claim the universe to be expanding and so, everything should be moving farther and farther apart; collisions, it would seem, should be highly improbable occurrences. However, this view is too simplistic and absorbing actions, such as that envisaged by black holes, are readily incorporated into Big Bang theory. The Arpian view of what is happening is taken to be in direct opposition to the Big Bang theory, probably because it may be interpreted as implying creation of matter and this notion is, of course, at the heart of the new quasi-steady state theory of Hoyle and his collaborators[*], as well as being seemingly contrary to well-established conservation laws. This quasi-steady state theory is a modification of the old steady state theory proposed by Bondi, Gold and Hoyle[†] many years ago and is a modification proposed in answer to criticisms of the original. It might be argued that they have listened to their critics and attempted to provide an answer. The difference between this modification and changes made to the Big Bang theory is that, in this case, it seems that the theory was modified but, in the case of the Big Bang, it seems that, when a problem is pointed out, something is simply added on in an attempt

---

[*] F. Hoyle, G. Burbidge & J. V. Narlikar, 2000, A Different Approach to Cosmology, (Cambridge U. P., Cambridge)
[†] H. Bondi & T. Gold, 1948, M. N. R. A. S., 108, 252. F. Hoyle, 1948, M. N. R. A. S., 108, 372

Exploding A Myth

to solve that immediate problem – other possible consequences of that possible solution are not always explored. However, once again the question of the role of scientific politics raises its ugly head and true science, which must be solely involved with a search for truth at all costs, seems displaced from the central position it should always occupy.

As indicated above, at one point in time - actually by about 1950 - there were really two rival theories attempting to explain the origin and workings of the Universe. These were the Big Bang and the Steady State Theory. Both accepted the idea that the Universe was homogeneous, isotropic and was expanding against the pull of gravity. However, the Steady State Theory assumed that matter could be both created and destroyed spontaneously, whereas the Big Bang did not. The idea of spontaneously creating or destroying matter challenges very widely, and strongly, held views on conservation and so will be anathema to many. On the other hand, one apparently awkward consequence of the Big Bang is that, at some time in the distant past, all matter seems to have been concentrated in some state of infinite density; that is, a singularity, the cosmic singularity, existed. It is often claimed that this singularity is a serious defect in the Big Bang theory on philosophical grounds but, in many areas of mathematics and physics, it is more usual to note that a singularity heralds the breakdown of a theory or that there are limits to the range of applicability of a particular theory. It is interesting to realise that, for some reason, no such restriction is imposed in this case

Big Bang Theory – Controversial or Not?

or, indeed, in the case of black holes of the type which are said to emerge via the general theory of relativity. In both these cases, attempts are actually made to give physical meaning to mathematical singularities. Apparently, it is this singularity in the case of the Big Bang which prompted Bondi, Gold and Hoyle to propose the Steady State Theory in which matter could be created spontaneously at a rate which compensated the reduction in density brought about by the cosmic expansion. Such a Universe would presumably have no beginning or end, it would have both an infinite past and future, but, possibly more importantly, the model would have no singularity.

Considering this latter point concerning the Steady State theory, it is interesting to wonder at the possible role played by fundamentalist religion in the seemingly widespread acceptance of the Big Bang and the resultant rejection of Steady State theory. A moment's reflection indicates that the possibility of such a link is not totally ludicrous. In chapter 1 of the first book of the Bible, *Genesis*, it is written:

"1. In the beginning God created the heaven and earth.

2. And the earth was without form, and void; and darkness was upon the face of the deep. And the Spirit of God moved upon the face of the waters.

3. And God said, Let there be light: and there was light

4. And God saw the light, that it was good: and God divided the light from the darkness.

Exploding A Myth

5. And God called the light Day, and the darkness he called Night. And the evening and the morning were the first day.

6. And God said, Let there be a firmament in the midst of the waters, and let it divide the waters from the waters.

7. And God made the firmament, and divided the waters which were under the firmament from the waters which were above the firmament: and it was so.

8. And God called the firmament Heaven. And the evening and the morning were the second day.

9. And God said, Let the waters under the heaven be gathered together unto one placc, and let the dry land appear: and it was so.

10. And God called the dry land Earth; and the gathering together of the waters called He Seas: and God saw that it was good.

11. And God said, Let the earth bring forth grass, the herb yielding seed, and the fruit tree yielding fruit after his kind, whose seed is in itself, upon the earth: and it was so.

12. And the earth brought forth grass, and herb yielding seed after his kind, and the tree yielding fruit, whose seed was in itself, after his kind: and God saw that it was good.

13. And the evening and the morning were the third day.

Big Bang Theory – Controversial or Not?

> 14. And God said, Let there be lights in the firmament of the heaven to divide the day from the night; and let them be signs, and for seasons, and for days, and years:
>
> 15. And let them be for lights in the firmament of the heaven to give light upon the earth: and it was so.
>
> 16. And God made two great lights; the greater light to rule the day, and the lesser light to rule the night: he made the stars also.
>
> 17. And God set them in the firmament of the heaven to give light upon the earth.
>
> 18. And to rule over the day and over the night, and to divide the light from the darkness: and God saw that it was good.
>
> 19. And the evening and the morning were the fourth day."

The chapter then continues with a representation of how life arrived on the earth, beginning with the arrival of the fish of the sea, followed by the beasts of the earth, and culminating with the arrival of man. It is interesting, incidentally, to note how the ordering in this very brief résumé of the final twelve verses of the chapter agrees so well with that of modern theories of evolution. As for the first nineteen verses, the first obstacle to be overcome is the unscientific language used. However, when that is done, it becomes immediately apparent that one valid interpretation of what appears in print is that the Universe was created

Exploding A Myth

quite suddenly, spontaneously in fact. The ordering that follows also links quite well with Big Bang philosophy. It might be argued, quite reasonably, that light would be necessary before grass and fruit trees could exist but, bearing in mind that the ideas, or stories, of *Genesis* are extremely old and may be interpreted sensibly only as representations produced by people without the benefit of modern scientific knowledge to illustrate, to a scientifically uneducated people, the beginnings of the Universe and of life on earth, the correspondence with the ordering of events according to the Big Bang theory is remarkably close. It might be noted specifically that even the presence of radiation before the formation of the stars may be inferred from verses fourteen to nineteen inclusive. However, was *Genesis* ever intended to be taken literally? Was it ever meant to be the literal truth describing the origin of the Universe and life in it? On this question, as with questions of theories of evolution, various views abound. Amongst these, is the view that the answers to the above two questions are in the affirmative. There are, and always have been, people who do believe the book of *Genesis* to be literally true. Some of these people are, and have been, serious scientists. This may seem almost a contradiction in terms but it is, nevertheless, true. It is, therefore, not difficult to see precisely how the Big Bang theory will appeal to such people as being the perceived 'Word of God'. It is very easy, but also very unfair, to ridicule such a standpoint, since the obvious temptation is so strong. Added to this is, of course, the additional realisation that at least one of the advocates of the Steady State Theory, Sir Fred Hoyle, was openly

Big Bang Theory – Controversial or Not?

something of a religious sceptic. It might be remembered that Hoyle's views on religion had caused something of a national uproar in Britain when he voiced at least some of them during one of his appearances on B.B.C. radio. It is even said that, at one point, the Astronomer Royal had to intervene when a sharp difference of opinion developed between the Chairman of the B.B.C. and the Archbishop of Canterbury over whether or not Hoyle should be allowed to continue to enjoy the freedom of the airwaves. Whether this final point had any influence on the way things turned out is not known for certain but, human nature being what it is, it is fairly easy to think it may have. This short semi-religious discussion merely serves to raise another question and that is whether, or not, religious fundamentalism played any part – however small – in the acceptance of the Big Bang Theory over the Steady State Theory? Indeed, it is not unreasonable to wonder if, with the seeming resurgence of religious fundamentalism in present day society, it is one factor keeping the Big Bang theory so much to the fore. It is still the case that the validity of the Big Bang theory seems accepted totally without question by much of the world-wide scientific community. A final point which might be remembered is, of course, that the creation story presented in the first chapter of *Genesis* is not peculiar to Judaism and Christianity. It is, for example, felt by some that the account appearing in *Genesis* actually originated in Babylonian theology. Whether or not this is true is not really of any importance in the present context, except to illustrate the fact that the story is definitely not a unique one and so, one must wonder at the worth of anyone assigning to it any undue authority. Also, as was pointed out by Stein Johansen, a similar creation story forms a part of Norse Mythology, adding further weight to the view that little reliance may be placed on the precise detail of these religious creation stories; their task is surely to illustrate how things might have originated and evolved, not to be regarded as literally true.

As a small additional point, it is interesting to note that, in a recent copy of *The Observatory* (vol.**125**, no. 1189, page 347), it was reported that an attender at a meeting of the Royal Astronomical Society had had a conversation with Tommy Gold in 1972 and the question of continuous creation had been raised. Apparently Gold claimed that "it was not that it was more elegant than the Big bang theory, but it was *not* the story of *Genesis*"". Again a possible link between Christian fundamentalism and science appears. It is not that the two should necessarily be totally separate from each other but the apparent nature of the link is what gives cause for concern. Put simply, is fundamentalist religion exercising an unhealthy influence over science in this area of cosmology, or anywhere else in science for that matter? It seems to be a question well worth raising and one of which serious scientists should be fully aware.

However, after this semi-religious digression, a return to the purely scientific arguments for the Big Bang is now necessary. As mentioned earlier, one of the most vociferous of early proponents of the Big Bang Theory was George Gamow. He and Ralph Alpher first put the theory forward seriously in 1948 and almost immediately became engaged in a war of words with the supporters of the Steady State Theory. However, Gamow's theory did, apparently, make some important predictions. Namely that there should be an abundance of helium of about twenty-five per cent by mass, and that it should be possible to observe the remnants of the radiation from the early hot phase of the universe's existence and this should be an isotropic radiation field with a black body spectrum with a temperature of a few degrees. The estimates forwarded for this temperature varied, however, between about five degrees and fifty degrees absolute. This was interesting because, as early as 1926, Sir Arthur Eddington[*] had predicted a temperature of space of three degrees absolute, purely on the basis of the radiation received from the stars. This calculation is very crude but the magnitude of the result provides food for thought, if nothing else. Again it might be noted that Eddington was discussing the temperature of interstellar space due to stars in our own galaxy; he was not considering intergalactic space. However, be that as it may, the Big Bang Theory received possibly its biggest boost, both within and without the scientific community, with the discovery by Penzias and Wilson of the cosmic background radiation, - that background radiation which is almost universally recognised nowadays as a left-over of the original Big Bang. Here it is important first to ask whether or not this discovery of the cosmic background radiation is, in fact, really due to Penzias and Wilson. It

---

[*] A. Eddington, 1988, The Internal Constitution of the Stars, Cambridge U. P., Cambridge.

Exploding A Myth

must be acknowledged that the existence of this background radiation was not universally recognised at the time of Penzias and Wilson. However, its existence had been detected in the late thirties and early forties by various astronomers. In 1941, McKellar had interpreted the observed data and had shown it to be caused by radiation excitation, which was taken to be black body and the temperature required for the observations to be properly explained was found to be 2.3°K. Hence, the detection of the microwave background should more correctly be dated from 1941. It is, in all fairness, understandable that this did not happen. In 1941, the world was in turmoil at the height of the Second World War and McKellar's important work did not appear in a front line journal. However, the truth has been known for some time now. Hoyle, in particular, has not been backward in publicising its existence. It is to be hoped that McKellar will soon be given the credit he surely deserves.

Actually, estimates of the temperature of intergalactic space go back as far as the end of the nineteenth century, at least, with the work of Guillaume who, like Eddington as mentioned above, was concerned with the temperature of interstellar space due to stars belonging to our own galaxy. It was in 1896 that Guillaume estimated this temperature to be between 5°K and 6°K, whereas, in 1926, Eddington, in his book *The Internal Constitution of the Stars*, estimated the temperature to be 3.18°K. It is of interest to note at this point that Hubble only established the existence of galaxies other than our own in 1924, so the fact that Eddington and

Big Bang Theory – Controversial or Not?

especially Guillaume confined their attentions to estimating the temperature of interstellar space due to stars in our own galaxy, rather than that of intergalactic space, is not surprising. However, it is possibly of interest to note that, nowadays, it is assumed that the visible universe has a radius of 10^{29} cm and contains 10^{11} galaxies, each composed of 10^{11} stars. If, following Eddington, it is assumed further that the heat received from the stars corresponds to that received if all the stars were of apparent bolometric magnitude 1.0, then, since each star of this absolute bolometric magnitude radiates 36.3 times as much energy as the sun, or 1.37×10^{35} ergs/sec, this leads to a figure of

$$1.09 \times 10^{-24} \text{ergs/sq cm/sec}$$

over a sphere of radius 10^{29} cm. The corresponding energy-density is obtained by dividing by the speed of propagation, which is 3.0×10^{10} cm/sec and this gives a figure of

$$3.63 \times 10^{-35} \text{ergs/cc.}$$

Accordingly, the total radiation of the stars is estimated to have an energy density of

$$3.63 \times 10^{-13} \text{ergs/cc.}$$

By using the well-known formula

$$E = aT^4,$$

where a is the coefficient in Stefan's law and has the value 7.64×10^{-15}, it follows that the temperature of space is

Exploding A Myth

$$T \cong 2.6°K$$

approximately. Obviously, this calculation is rather crude but the end result is interesting and seems to indicate that another explanation for the temperature of space, other than attributing it unquestioningly to the results of the Big Bang, is at least feasible.

It might usefully be noted that, long before Gamow and others began to espouse the Big Bang theory, several notable scientists had followed the lead of Guillaume and Eddington and proposed estimates of the temperature of intergalactic space. Following initial work by Millikan and Cameron in which it was deduced that the total energy of cosmic rays at the top of the atmosphere was a tenth of that due to the heat and light emitted by the fixed stars, Regener eventually concluded that both energy fluxes should possess more or less the same value. In an article of 1933, he used this as a basis for deducing a value of 2.8°K as the temperature characteristic of intergalactic space. This work was discussed favourably by no less a person than Walther Nernst who, by 1912 had developed the notion of a stationary state universe. By 1937, he had further developed this and actually proposed a 'tired light' explanation for the cosmological redshift; that is, he suggested that absorption of radiation by an aether which decreased the energy and frequency of galactic light. Whether one accepts or rejects these ideas now, it should be noted that, in all these separate pieces of work, as well as in subsequent examinations by such as Max Born, Stefan-Boltzmann's law, which is characteristic of black body radiation, is of paramount

Big Bang Theory – Controversial or Not?

importance. Also, in none of this work, nor in that of McKellar, is any reference made to the Big Bang theory; it is simply not necessary to introduce it in order to achieve the results cited!

However, nowadays it is the papers by Gamow and by Alpher and Herman, dating from 1948[*], which tend to hold pride of place where discussion of the background temperature is concerned. They pointed out that, if helium was synthesised in the early universe, then, in present times, there should exist a radiation field with a temperature of approximately 5°K. Gamow offered another prediction of the temperature of the background radiation in his 1952 book *The Creation of the Universe*, but this time the estimate, which was claimed to be "in reasonable agreement with the actual temperature of interstellar space", was roughly 50°K. Nevertheless, it is frequently claimed that Gamow and his collaborators predicted the 2.7°K temperature (even though their lowest estimate was in fact 5°K) before the 'discovery' of Penzias and Wilson, whereas the steady state theory did not. This was, and is still, hailed as one of the strongest arguments in favour of the Big Bang. However, it must not be forgotten that the original steady state theory did not rule out the existence of a background radiation and, as is pointed out in Hoyle's last book, some unpublished calculations by Hoyle, Bondi and Gold, dating from about 1955, indicated a temperature associated with that radiation of 2.78°K. Obviously, revealing this at the time of the supposed

---

[*] G. Gamow, 1946, Phys. Rev., 70, 572

Exploding A Myth

discovery of the cosmic background radiation would have produced totally the wrong reaction.

In his later years, Hoyle and his collaborators produced a modified form of the steady state theory – the so-called quasi-steady state cosmology - in an attempt to answer their critics and to restore some much needed open-minded debate to this major question of the origin of the universe. In a truly open-minded scientific world, this new theory would be viewed afresh and without preconceived notions being allowed to dominate, but will it receive that fair hearing? However, in the present context, it is probably far more important to note that, from the outset, the Steady State Theory never ruled out the possibility of there being a background radiation in existence. Therefore, it is obviously totally incorrect to use the existence of this background radiation as a major reason for attempting to discount the Steady State Theory – original or modified!

When introducing the articles by Gamow and by Alpher and Herman above, it was noted that they made reference to the synthesis of helium in the early universe. They were using this to support their claim that, if this were so, a radiation field pervading the whole of space should exist now. This, of course, raises the entire question of the process behind the synthesis of helium and the other light elements. It is of interest to realise that, once again, the papers referred to here were not the earliest attempts to raise this problem. Actually, the earliest article by Gamow appeared in 1946[*]. In it, he argued that, in the early universe, the chemical elements were synthesised by neutron addition. Hoyle also produced his first article on stellar nucleosynthesis in 1946[†] and, interestingly, his view was the direct opposite of that proffered by Gamow. In fact, it is quite widely accepted now that the originator of the theory behind the synthesis of the light elements was Hoyle and a great many people are still puzzled by the fact that he received no part of the Nobel Prize awarded for that work. While the overall thesis of this present work is concerned with the place of accepted 'conventional wisdom' in the scientific world, this treatment of Hoyle inevitably raises the spectre of 'politics' within the scientific establishment. However, now return to the articles by Gamow and by Alpher and Herman. After one or two early hiccups, Gamow and his collaborators produced a theory whose key point was the essential requirement that an amount of helium be synthesised in agreement with the observed value of approximately 0.25 by mass when compared with hydrogen. It might be noted immediately that this fraction is not thought to be constant in time and that alone raises questions. Although it is known that helium is produced from hydrogen in the interior of stars, it was always felt, and still is, that stellar synthesis would make only a negligible contribution to this observed fraction. As has been pointed out by Hoyle and his collaborators, it would take of the order of 10^{11} years to increase the

---

[*] G. Gamow, 1948, Phys. Rev., 74, 505. R. A. Alpher & Herman, R., Phys. Rev., 75, 1084

[†] F. Hoyle, 1946, M.N.R.A.S., 106, 343

value of this fraction from zero to 0.25. Around 1950, when these initial calculations were instigated, the Hubble constant was believed to hold a value leading to the age of the universe being only of the order of 10^9 years. Since this figure was so much less than the time apparently required for the mass fraction of helium to be explainable from astrophysical processes, it was decided that it needed to be explained via primordial synthesis in the very early universe. The first crucial realisation to follow this decision was that it could be true only if the energy density of radiation in the early universe was large compared with the rest mass energy of matter. Accepting this was a major change in thinking for many since, up to that point, the opposite had been assumed true. An immediate consequence was that the radiation temperature had to be inversely proportional to the square root of the time. Up to this point, the argument was not unreasonable given the initial assumptions but what followed was a completely ad hoc step and it should be noted that it remains *ad hoc* today. The mass density of stable non-relativistic particles – neutrons and protons – decreases with the expansion of the universe and Alpher and Herman denoted this by ρ and took

$$\rho = 1.7 \times 10^{-2} t^{-3/2} \text{ g/cm}^3.$$

Here it is the choice of the coefficient of proportionality as 1.7×10^{-2} which is the *ad hoc* step. There is absolutely nothing in the theory of the Big Bang which actually fixes the value of this coefficient. It is a choice made quite freely but a choice which has the enormous, but to many acceptable, effect of ensuring that the

Big Bang Theory – Controversial or Not?

calculated value for the mass fraction of helium is indeed 0.25, in accordance with the observed value. This must mean, however, that as the value of the said mass fraction changes, as it surely must over an extended period of time, the value of this constant of proportionality must change also. The obvious question to follow then is, does the Big Bang theory, therefore, actually *predict* the correct value for the mass fraction of helium? The answer has to be an emphatic 'No'!

It is, unfortunately, true to note that often, at the end of their undergraduate days, many students emerge totally convinced that the big bang theory correctly describes the beginnings of our universe and also many of its subsequently developed properties. They believe it to be the only theory which explains the cosmic microwave background radiation; they believe it to be the only theory to explain the mass fraction of helium. This, and much more, has all been learnt in undergraduate courses as being absolutely sacrosanct. Further, these beliefs are vigorously supported by so many popular science books, such as Simon Singh's *Big Bang*, and by many popular science lectures. The popular science lecture on the big bang by Simon Singh, which has received quite widespread publicity, is an excellent example. This lecture is beautifully constructed and presented, as one might expect from an experienced member of the BBC personnel able to call on the resources of that organisation if necessary. The personality presenting the talk is friendly and engaging; a young audience, in particular, is rapidly enthralled. The use of power point to deliver the message, and message it is, is very professional. Everything about the talk from a delivery viewpoint is beyond reproach, and that is where the danger lies. Young people with impressionable minds will leave such a talk totally convinced that they have just been exposed to an enunciation of the complete truth regarding the birth of our universe. But have they? They will have been told, amongst other things, that the cosmic background radiation was discovered by Penzias and Wilson in 1965. McKellar's work will have been ignored. The steady state theory will have been dismissed totally with hardly a glance in its direction and no mention will have been made of the newer modified theory. The constant need to add to, and modify, the original Big Bang theory with entities such as dark matter and dark energy – a topic to be discussed further a little later - will have been glossed over. However, in the case being highlighted here, the presentation will have been so slick and professional that these points will not have sunk in to members of the audience. Many of the enthralled youngsters will probably leave the lecture theatre remembering more that Simon Singh would like to be admired by Cameron Diaz in the same way that Einstein was apparently admired by Tallulah Bankhead, than that they have just heard details of *one* theory for the beginning of our universe. Superficial gloss will have prevailed. As stated previously, herein lies the danger. The scientists of tomorrow are not being trained to have open questioning minds. Rather they are having their minds programmed to be closed to all thoughts which might possibly conflict with 'conventional wisdom'. The message often appears to be delivered

Big Bang Theory – Controversial or Not?

with what amounts to an almost religious fervour, – what might be termed scientific evangelism.

It is possibly of interest to note that, once again, religious vocabulary seems to be occurring naturally in the attempt to explain attitudes within certain areas of science. No doubt, some scientific practitioners would be appalled at this and more especially with any thought of there being any possible analogy linking science and religion, however abstrusely. Nevertheless, such an analogy concerning attitudes and approach appears to be an ever returning theme. The possibility of a linguistic link, if nothing else, between the Big Bang theory and the creation story as presented in the first chapter of the *Book of Genesis* is clear for all to see. The fact that Fred Hoyle became unpopular in certain circles because of his atheistic views, rather publicly aired on BBC radio, is well-known. Is there a link between the two? Were, in fact, both issues in the popular acceptance of the Big Bang theory as opposed to other theories, in particular the original Steady State theory? After all, it must be remembered that the Steady State theory is still summarily dismissed as a serious attempt to explain the universe in which we exist. However, at this point in time, it should be noted that, even without the latest modifications to the theory, the advocates of Steady State had answered many of the criticisms of that theory quite convincingly. The whole history of what Hoyle and his associates term 'the war of the source counts' provides a classic example of this. The details of this controversy are well documented, by those deeply involved on one side of the argument, in *A*

Exploding A Myth

Different Approach to Cosmology by Hoyle, Burbidge and Narlikar[*]. Here it is discussed in detail how, initially, it appeared that the Steady State Theory indicated incorrect results when it came to examining radio sources and their distribution. Essentially, it seemed that the data collected allowed either of two possible conclusions to be drawn. Ryle and his collaborators at Cambridge took one view; Hoyle subscribed to the alternative. This meant that Ryle and his supporters viewed the data in a way which opposed the validity of the steady state theory. The argument certainly raged fast and furious for many years but, in the end, following queries raised by Robert Hanbury-Brown at the Paris Symposium as early as 1958, the truth finally emerged following work published in 1988. In truth, some objections to the original Steady State theory were destroyed at this point. However, this occurred some thirty years after the queries first erupted onto the scientific scene. Too much time had elapsed; too many opinions had been irrevocably formed; there was little or no chance that any change in popular scientific opinion would be accomplished. The modified theory, presented so eloquently in the above-mentioned book, is also not likely to create a revolution in scientific thought on this matter, - at least not immediately. Positions are far too entrenched; too much 'face' – and, possibly more importantly, too many positions of power and influence – would be lost if any senior scientist completed a *volte-face* on this issue. It is also sad to realise that many have been deterred from

---

[*] F. Hoyle, G. Burbidge & J. V. Narlikar, 2000, A Different Approach to Cosmology, (Cambridge U. P., Cambridge)

Big Bang Theory – Controversial or Not?

studying the Steady State theory because it is felt by so many to have been disproved by observations and, therefore, merits no further study. On the face of it, this is a not unreasonable stand-point, but no-one can claim seriously that there is a single undisputed theory describing all aspects of our universe and its origins. True the Big Bang theory seems, in some ways, the most successful theory so far but, at best, that is all that it is, - the most successful theory so far. In all aspects of science, practitioners should remain open-minded and, in this particular area, more so probably than in others. It is incumbent on all – amateur as well as professional – to keep all options open and that means remaining fully up-to-date and conversant with all of the modified Steady State theory, as well as the present version of the Big Bang.

However, to return to the actual Big Bang theory, a further problem faced by the adherents to the theory is the seemingly constant need to add to the basic theory in order to overcome problems. Obvious examples of this are the introduction of the ideas of inflation, dark matter and even dark energy. It is, however, the first of these additions to which attention must be turned. The Big Bang model was faced with the 'horizon' and 'flatness' problems. The first of these arises from the prediction that the Universe is both homogeneous and isotropic, which implies that, in the early Universe, disconnected regions would have had to have been in nearly the same state to achieve the present-day homogeneity. The lack of contact makes such a scenario extremely unlikely. The second paradox concerns the

Exploding A Myth

extrapolation of the present value of the ratio of the energy density of the Universe to the critical energy density back to the Big Bang. When this is done, the extremely unlikely value of nearly unity is found. In 1981, Guth[1*] attempted to address these by releasing the assumption of the adiabaticity of the early expansion of the Universe. This resulted in the so-called inflationary scenario, which supposes that a supercooling of the material of the Universe led to a period of exponential growth involving the release of the latent heat of the phase transition and an increase in the entropy of the Universe. Modifications to this basic model were introduced by Linde[2] and Hawking and Moss[3] to attempt to overcome the fact that it would produce large inhomogeneities which are incompatible with observation. The exponential dependence of the scale factor on the time is certainly a solution of the equations of general relativity, but the association of the release of a latent heat is not. This central objection went unnoticed until recently[4].

The Einstein equations resulting from the Robertson-Walker metric are:

$$\ddot{R} = -\frac{4\pi G}{3}(\varepsilon + 3p)R$$

and

---

[*] A. Guth, 1981, Phys. Rev. D, 23, 347
[2] A. D. Linde, 1982, Phys. Lett. B, 108, 389
[3] S. W. Hawking & I. G. Moss, 1982, Phys. Lett. B, 110, 35
[4] B. H. Lavenda & J. Dunning-Davies, 1992, Found. Phys. Lett. 5, 191

$$\left[\frac{\dot{R}}{R}\right]^2 + \frac{k}{R^2} = \frac{8\pi G \varepsilon}{3},$$

where $k = +1$, -1, or 0 depending on whether the Universe is curved ($k = +1$ or -1) or flat ($k = 0$) respectively. If the second of the above equations is differentiated with respect to t and the second derivative eliminated,

$$\frac{d}{dt}(\varepsilon R^3) + p\frac{d}{dt}(R^3) = 0$$

results, where ε is the energy density and p the pressure. Comparing this with the well-known thermodynamic result

$$Td(sR^3) = d(\varepsilon R^3) + pd(R^3),$$

where s is the entropy density and sR^3 the total entropy in a volume whose radius of curvature is R, shows that Einstein's equations imply adiabaticity:

$$d(sR^3) = 0.$$

Hence, no criterion for non-adiabatic growth can arise from Einstein's equations.

An expanding Universe, as suggested by Hubble's observation of galactic expansion, will involve progressively increasing compression in the past. All that the inflation hypothesis was designed to do would be achieved by a speed of light which increases with increasing temperature, as was mentioned earlier when discussing the television

Exploding A Myth

programme *Einstein's Biggest Blunder*. Of course, this alternative description of the past is not compatible with the universal application of the principle of general relativity which requires a universal speed for light.

It is not without interest to realise that additions to the Big Bang theory are accepted unerringly. Seemingly, no questions are raised when these new notions such as inflation, dark matter and dark energy are introduced in attempts to preserve this theory as the only acceptable explanation for our universe as we see it. However, there doesn't appear to have been any significant upsurge in interest in the Steady State Theory since the publication of all the material – both strictly academic and semi-popular – advocating modifications to that theory. Many will claim this due to the fact that the theory is quite simply incorrect, but the facts don't support this view. Neither do they support the view that the Big Bang theory is true beyond all reasonable doubt. In reality, the truth must lie either somewhere between these two extremes or possibly completely outside these two interesting attempts buried in some, as yet, totally unknown theory. We really truly understand very little, however great mankind's scientific achievements may be thought to be. When we understand in detail what is meant by terms such as 'force' and 'mass', then we will be on the way to a complete understanding of our universe and all that exists in it but, until that time, it seems sensible to retain all options and that must include both the Big Bang and the Steady State theory, together with any other thoughts, as possible explanations. Prominent

Big Bang Theory – Controversial or Not?

among these other thoughts must be the so-called 'tired light' theory. So much in our presently accepted theories depends on the interpretation of the red-shift phenomenon. It is commonly accepted, as has been mentioned already, that this red-shift is brought about by the Doppler shift of light due to the recession of distant galaxies. However, at least theoretically, other explanations are feasible. A brief outline of the worries expressed by Halton Arp has been discussed earlier. However, another possible explanation for the existence of the observed red-shifts is provided by the notion of 'tired light'. Here the basic idea is that quanta of light could actually lose energy during their journey through space from distant galaxies to us. The suggested decrease in photon energy would result in an increase in wavelength that would be proportional to the distance travelled. This would, of course, be viewed as a reddening. Another contributory factor to this reddening of light could be provided by scattering by particles of intergalactic dust. Probably the effect of scattering by dust particles may be discounted at this stage, though not entirely forgotten, because such scattering would be expected to result in a broadening of the spectral lines and that is not observed. However, the general notion of 'tired light', while dismissed almost out of hand by most workers in the field, cannot be totally abandoned as yet. Firstly, it is a theory which has a long history and which has never gone away completely. It has been advanced and supported by a powerful array of physicists from Max Born to Jean-Pierre Vigier. This, in itself, is not sufficient to make the theory acceptable, but it is surely a good enough reason for it to be taken

Exploding A Myth

seriously. Some wish to dismiss it on the grounds that only in Big Bang cosmology is there a satisfactory explanation provided for the origin of the cosmic background radiation and for the abundance of the light elements. However, as has been seen already, this is simply not true. The case of the Steady State Theory proves this beyond reasonable doubt. Whether one believes or disbelieves the Steady State Theory or, for that matter, the Big Bang theory, it is certainly true to say that, in attempting to destroy the Steady State Theory, the truth was not to the fore. It is disturbing to realise that this explanation is the one advanced for dismissing so many suggestions and it is no more true today than it was when first put forward and agreed. 'Tired light' may not be a true explanation for any of the questions arising in cosmology but, like anything else, it deserves to be viewed with a completely open mind before a decision is reached. Once again it is seen that this is the true problem facing cosmology as a whole and the Big Bang theory in particular – both must be viewed and assessed with a completely open mind. Personal preferences and prejudices have no place, no place at all, in the evaluation of a scientific theory. The task must be accomplished purely by using the accepted methods of science and known scientific knowledge - always realising, of course, that any conclusion will be subject to limitations placed on its validity by the extent of such knowledge at any one time.

A further major problem facing this area is associated with the advance of knowledge. In this colossal area, knowledge advances through careful, painstaking observation of the cosmos. All the observations made must then be processed most carefully. This again is something which is not quite so straightforward as might appear at first. Quite frequently, data has to be analysed statistically and it is crucial that this is done completely honestly. There must never ever be even a suspicion that an effect is claimed which might be simply due to the statistical package used for the analysis. Hence, this again is something which must be undertaken by truly open-minded people and making use of professional statisticians to analyse data – rather than it being done by those who might be thought to have a vested interest in the end result – could be a sensible way forward in this area. Too often the impression is left that the conclusion announced is merely confirmation of the result 'expected' before the experiment or observation was begun. In a way, this brings a return to the case of Halton Arp. As has been noted earlier, many astronomers are said to doubt Arp's interpretation of the photographs he has taken and usually their scepticism is said to be based on some aspect of the statistical analysis of his data. It has been claimed, though, that if a continuous change in red-shift values could be measured along an apparently material link between a low red-shift galaxy and a high red-shift quasar, then Arp's view would be vindicated. However, it seems that no such effect has been found as yet, although strenuous efforts are said to have been made to establish the presence, or absence, of such an effect. This again raises the question of whether or not observers are finding what they want to find rather than the truth. Some ask at this point, 'What is truth?' No

Exploding A Myth

doubt a deep philosophical discussion could ensue here. However, suffice it to note that the Oxford Dictionary states that one meaning of the word 'true' is "in accordance with fact or reality, not false or erroneous". It goes on to state that 'truth' is the "quality, state, of being true". These elementary definitions of the two words give a clear everyday meaning of what they mean in the present circumstances. Indulging in philosophical discussions surrounding the meanings of words doesn't necessarily help anyone; it frequently serves simply to divert attention from the question at issue, - in this case that of the major problems facing science today. As with so many of the major controversies in science, positions have become entrenched, 'conventional wisdom' has become almost indelibly etched into the folklore surrounding the subject. Young scientists are, all too often, taught established truth as if it were religious dogma. They are not trained to really think; only to think along well-established lines – lines drawn by the 'Gods' of 'conventional wisdom'. This probably sounds harsh and seemingly linking science with religion again will undoubtedly offend some who feel the two separated by an infinite chasm. It will probably offend others, like Dawkins, also, who claim that scientific truth is paramount and using its clearly defined techniques leads to a 'proof' that no God exists. Unfortunately, the truth often does hurt and, in reality, young scientists are all too often indoctrinated with supposed 'facts', rather than educated to have open, enquiring minds. If the result of raising these unpleasant aspects of present day world science is to reintroduce an

Big Bang Theory – Controversial or Not?

open questioning attitude into science, then the imagined hurt will have been more than worthwhile.

As an addendum to this discussion of the Big Bang, it might be noted that an entire edition of the well-known and well-respected British Broadcasting Corporation's television science programme, Horizon, was devoted to the present-day search for dark matter[*]. The programme title was *Most of our Universe is Missing*; an eye-catching title guaranteed to attract viewers. It pointed out that some scientists feel it not known from what much of our universe is made; others argue that some presently accepted theories, such as Newton's law of gravitation, may be wrong – or, at least, only apply locally rather than globally. The programme itself contained much of genuine, but not probably general, interest. However, one worrying aspect in the present context was the fervour exhibited by several contributors in support of the Big Bang as explaining the origins of the universe. Only one really drew back to point out that the Big Bang is a theory, and only a theory! As was asked in a recent letter to *The Observatory*[†], "When will the *Cosmological Establishment* stop calling their theory the truth, the whole truth, and nothing but the truth?" Considering this assertion, it might be noted that, in his book *Before the Beginning*[‡], the Astronomer Royal, Sir Martin Rees, confidently states on page one that "Our universe sprouted from an initial event, the 'big bang' or

---

[*] Horizon, B.B.C, 9th.February, 2006

[†] A. Welch, 2006, *The Observatory*, **126** (no.1190), 51

[‡] M. Rees, 2002, *Before the Beginning*, (The Free Press, London)

Exploding A Myth

'fireball'" - a very bold, categorical statement with which to start any account, but is it really true? Anyone who questions it is said to belong to a minority. Apparently, most cosmologists would offer strong odds on there having been a 'big bang', feeling that "everything in our observable universe started as a compressed fireball, far hotter than the centre of the Sun". The idea that this scenario is questioned by a minority only would seem true, but largely because so many in science possibly feel it in their own best personal interests to conform to the imposed dictats of 'conventional wisdom'. Of course, in these terms, that 'minority' might really be a 'silent majority'. As for those outside professional scientific circles, those who in the final analysis pay the bills, they have been subject to so much publicity, via all media forms, in favour of this theory to the exclusion of all else, that it is no wonder they believe it to be an unassailable truth, not simply a mere theory. However, as another contributor to *The Observatory* pointed out[*], because the Steady State theory appears to provide precise predictions, it seems to have suffered in comparison with other theories, such as the Big Bang, which allow scope for empirical adjustment. This writer felt it precisely this which made the steady state theory a good theory and seemed to feel it likely that that theory would return eventually in some form. Be that as it may, it is undoubtedly of interest to speculate on what the future holds in this field, but one thing is absolutely certain, for real progress to be made, investigators must retain

---

[*] P. Fellgett, 2006, *The Observatory*, **126** (no.1190), 51

open minds; very little should be totally discarded at this juncture. In the present atmosphere that seems a lot to ask, but it is absolutely essential if science is to advance positively!

Chapter Four

The Schwarzschild Solution and Black Holes

It was John Michell[*] who, in 1784, first derived an expression, using Newtonian mechanics, for the ratio of the mass to the radius of a spherical body having an escape speed equal to, or greater than, the speed of light. It should be noted here that, although the idea of light possessing a finite speed was known at Michell's time, that speed was not regarded as an ultimate speed. The notion of an ultimate speed, if indeed such a speed truly exists in nature, only surfaced with the emergence of the special theory of relativity. However, it was towards the middle of the last century that the modern idea of a relativistic black hole appeared. This latter object actually arose as a physical explanation of a singularity apparently occurring in the Schwarzschild solution to the field equations of general relativity. It is interesting to note that this singularity occurs when the ratio of the mass to the radius (or, in this case, the radius of the so-called event horizon) formally satisfies the same relation as that deduced by Michell. This idea of a 'black hole' (but probably better termed a dark body), a body from which nothing can escape – not

---

[*] J. Michell, 1784, Philos. Trans. R. Soc., 74, 35

even light - has proved an extremely popular topic of uninformed conversation, and has become especially beloved by science fiction writers. However, the modern notion, as distinct from the original idea of Michell, faces several problems. Possibly the most problematic is the fact that, in Schwarzschild's original article[*], this crucial singularity does not appear. In fact, the form of the 'Schwarzschild solution' appearing in so many texts is one resulting from use of a co-ordinate system different from the spherical polar co-ordinates so meticulously used by Schwarzschild himself. Hence, the crucial singularity is completely dependent for its existence on the system of co-ordinates used and so cannot possibly have any physical significance assigned to it; it is purely a product of the co-ordinate system adopted. A further problem facing the idea is simply that Einstein himself, often referred to by some as the 'father of black holes', went to great lengths, in an article of 1939[†], to show that the mentioned singularity had no physical significance. It is interesting to note that neither Einstein nor Schwarzschild claimed the offending singularity to have any physical significance, but their strongly held opinions have been over-ruled to the extent that, somewhat ironically, not only is Einstein credited with being the father of black holes, but the uncharged, non-rotating black hole is commonly termed a Schwarzschild black hole.

---

[*] K. Schwarzschild, 1916, Sitzungsberichte der Königlich Preussischen Akademie der Wissenschaften zu Berlin, Phys.-Math. Klasse, 189(translation by S.Antoci & A.Loinger, arXiv:physics/9905030)
[†] A. Einstein, 1939, Annals of Mathematics, 40, 922

Exploding A Myth

However, referring back for a moment to Michell's spherical body with an escape speed greater than or equal to the speed of light. For such a body, the ratio of its mass to its radius would have to be greater than or equal to 6.7×10^{26} kg/m. For the relativistic black hole, the expression would be exactly the same but, instead of radius, it is the radius of the event horizon that would be being considered. Regularly these days one reads of black holes being identified positively. Most, if not all, galaxies are supposed to possess a massive central black hole. However, as yet, no black hole has been identified beyond reasonable doubt. In no case has the ratio of mass to radius satisfied the mentioned inequality and what some regard as the defining feature of a black hole – its event horizon – has never been positively identified. All the evidence to support the existence of these supposed black holes has been circumstantial. Most importantly also, the starting point for discussing some observational data has been the 'fact' that a black hole does exist at the centre of a particular galaxy or just simply that black holes do exist and so do offer valid explanations of data. Black holes may exist but, if they do, they will surely emerge naturally out of some more complete theory of stellar evolution than exists at present, rather than as the rather dubious consequence of attempting to impose a physical explanation on a mathematical singularity.

In many of the standard textbooks on the General Theory of Relativity[*], time is devoted to discussing

---

[*] R. Adler, M. Bazin, & M. Schiffer, 1965, Introduction to General Relativity, (McGraw-Hill, New York)

The Schwarzschild Solution and Black Holes

Schwarzschild's solution of the Einstein field equations. Normally, this solution is stated as being either

$$ds^2 = \left\{1 - \frac{2Gm}{rc^2}\right\}c^2 dt^2 - \left\{1 - \frac{2Gm}{rc^2}\right\}^{-1} dr^2 - r^2\left(d\theta^2 + \sin^2\theta d\phi^2\right)$$

(1)

or more usually

$$ds^2 = \left\{1 - \frac{2m}{r}\right\}dt^2 - \left\{1 - \frac{2m}{r}\right\}^{-1} dr^2 - r^2\left(d\theta^2 + \sin^2\theta d\phi^2\right)$$

(2)

where the universal constant of gravitation, G, and the speed of light, c, have both been put equal to unity. Here r, θ, and ϕ appear to be taken to be normal polar co-ordinates.

In the above expressions, a mathematical singularity is seen to occur when $r = 0$, as might be expected for polar co-ordinates. However, due to the form of the coefficient of dr^2, it follows that a second mathematical singularity occurs when, in (1), $rc^2 = 2Gm$ or, in (2), $r = 2m$. The first singularity is regularly dismissed as being merely a property of polar co-ordinates and, therefore, of no physical significance. The second singularity, however, tends to have a physical interpretation attributed to it - namely that it is said to indicate the existence of a black hole. Somewhat ironically, as will be seen later, this is referred to as a Schwarzschild black hole. If this interpretation were valid, it would

Exploding A Myth

imply that, for an object of mass m and radius r to be a black hole, it would need to satisfy the inequality

$$m/r \geq c^2/2G = 6.7 \times 10^{26} \text{ kg/m} \qquad (3)$$

As stated above, many modern texts quote one of equations (1) or (2) as the Schwarzschild solution of the Einstein field equations, but is this so? Recently, an English translation of Schwarzschild's article of 1916[*] has appeared and this has made the original work accessible to many more people. For this the scientific community owes the translators, S. Antoci and A. Loinger, a tremendous debt of gratitude. It also enables the above question to be raised by more people.

An excellent discussion of the Schwarzschild solution and its derivation is provided in chapter eighteen of the little book on the General Theory of Relativity by Dirac[†]. Here it is presented in the form (2) above and r, θ and ϕ are quite clearly stated to be the usual polar co-ordinates. It is pointed out that the case being considered is that of a static, spherically symmetric field produced by a spherically symmetric body at rest. After the completion of the derivation, it is noted that the said solution holds only outside the surface of the body producing the field, where there is

---

[*] K. Schwarzschild, 1916, Sitzungsberichte der Königlich Preussischen Akademie der Wissenschaften zu Berlin, Phys.-Math. Klasse, 189(translation by S.Antoci & A.Loinger, arXiv:physics/9905030)

[†] P.A.M. Dirac, 1996, General Theory of Relativity, (Princeton University Press, Princeton, New Jersey)

The Schwarzschild Solution and Black Holes

no matter and, hence, it holds fairly accurately outside the surface of a star.

The following chapter is then devoted to the topic of black holes. It is noted that the Schwarzschild solution (2) becomes singular when $r = 2m$ and so it might appear that that value for r indicated a minimum radius for a body of mass m but it is claimed that a closer investigation reveals that this is not so. In the discussion which follows, the continuation of the Schwarzschild solution for values of $r < 2m$ is investigated. To achieve this, it is found necessary to use a non-static system of co-ordinates so that components of the metric tensor may vary with the time co-ordinate. This is accomplished by retaining θ and ϕ as co-ordinates but, instead of t and r, using τ and ρ defined by

$$\tau = t + f(r) \text{ and } \rho = t + g(r), \qquad (4)$$

where the functions f and g are at the disposal of the investigator.

It transpires that, for the region $r < 2m$, the Schwarzschild solution is found to adopt the form

$$ds^2 = d\tau^2 - \frac{2m}{\mu(\rho - \tau)^{2/3}} d\rho^2 - \mu^2(\rho - \tau)^{4/3}\left(d\theta^2 + \sin^2\theta d\phi^2\right),$$

(5)

where $\mu = \left(\frac{3}{2}\sqrt{2m}\right)^{2/3}$.

Exploding A Myth

From the actual derivation, it follows that the critical value $r = 2m$ corresponds to $\rho - \tau = 4m/3$ and there is no singularity at this point in this metric.

From this point onwards, Dirac's argument becomes extremely interesting. He notes that the metric given by (5) satisfies Einstein's equations for empty space in the region $r > 2m$ because it may be transformed into the Schwarzschild solution by a simple change of co-ordinates. By analytic continuation, it is seen to satisfy the equations for $r \leq 2m$ also, because there is now no singularity at $r = 2m$. The singularity now appears, via equations (4), in the connection between old and new co-ordinates. Dirac then comments that, once the new co-ordinate system is established, the old one may be ignored and then the singularity appears no longer.

He comments further that the region of space for which $r > 2m$ may not communicate with that for which $r < 2m$. Also, any signal, even a light signal, would take an infinite time to cross the boundary at $r = 2m$. Thus, there can be no direct observational knowledge of the region for which $r < 2m$. If this argument were true, surely the region for which $r < 2m$ would lie outside our universe; would not really be a part of it? Dirac calls the region for which $r < 2m$ a black hole, but is this an object in our physical three-dimensional space or one in an abstract, four-dimensional, mathematical space-time?

Finally, Dirac asks whether such a region exists and notes that the only definite statement which may be

The Schwarzschild Solution and Black Holes

made is that the Einstein equations allow it. This is a question which will be considered further shortly but suffice it to say at this juncture that Einstein himself did not accept that it existed physically[*]. It is noted that a massive stellar object may collapse to an extremely small radius where the forces of gravity might become so strong that no known physical forces could withstand them and prevent further collapse. Such a situation would herald the collapse to a black hole but, as measured by our clocks, the final state would be achieved only after an infinite time. This argument would appear to stem from the ideas of Oppenheimer and Snyder[†]. They predicted that, when all sources of thermonuclear energy were exhausted, a large enough star would collapse and the contraction would continue indefinitely unless the star was able to reduce its mass sufficiently by some means. They also made the point that the total time for such a collapse would be finite for an observer co-moving with the stellar matter, although it would appear to take an infinite time for a distant observer. This was taken to indicate that the star tended to "close itself off from any communication with a distant observer"; only its gravitational field persisting. Accepting this argument as valid for the moment, it might be asked, if such an object existed, would it *ever* be detectable by an external observer? On the other hand, if its gravitational field persists, and presumably the effects of that gravitational field on the

---

[*] A. Einstein, 1939, Annals of Mathematics, 40, 922
[†] J.R. Oppenheimer, & H. Snyder, 1939, Phys. Rev. 56, 455

surroundings, then, in a sense, the star is retaining some contact, albeit indirect, with a distant observer.

Also, for very many years, it has been noted that the transformation
$$\tau = t + u + 2m\log(r - 2m)$$

applied to the Schwarzschild solution in the form (2) would remove the offending singularity. This was taken to indicate that the singularity was mathematical, *not* physical. This conclusion agrees with that of Einstein himself who, in an article of 1939[*], concluded that the result of the investigation contained in that paper was a "clear understanding as to why the 'Schwarzschild singularities' do not exist in physical reality". He went on to point out that, his investigation dealt only with clusters whose particles moved along circular paths but he felt it not unreasonable to feel that more general cases would have analogous results. He then stated quite categorically that "the 'Schwarzschild singularity' does not appear for the reason that matter cannot be concentrated arbitrarily". This seems a very definite rejection of the notion of black holes by the very man often heralded as their father. If the general tone of his book is an indication of his view, then it seems to be the case that Dirac agreed with this interpretation also. This point concerning a possible physical interpretation of a mathematical singularity has been raised previously by Loinger[†], who has published a number of articles on

---

[*] A. Einstein, 1939, Annals of Mathematics, 40, 922
[†] A. Loinger, arXiv:physics/0402088

The Schwarzschild Solution and Black Holes

arXiv.org in which the non-existence of black holes has been claimed. However, what of Schwarzschild himself? It's his solution of Einstein's equations which is really at the heart of this matter.

As noted earlier, the translation of Schwarzschild's paper of 1916[*] into English has made his work accessible to many more people. In his article, everything is written initially in terms of variables denoted by x_1, x_2, x_3, x_4 and the point is made that the field equations "have the fundamental property that they preserve their form under the substitution of other arbitrary variables as long as the determinant of the substitution equals one". The first three of the above co-ordinates are then taken to stand for rectangular co-ordinates, and the fourth is taken to be time. If these are denoted by x, y, z, and t the most general acceptable line element is then stated, but it is noted immediately that, when one goes over to polar co-ordinates according to the usual rules, the determinant of the transformation is not one. Hence, the field equations would not remain unaltered. Schwarzschild then employs the trick of putting

$$x_1 = r^3/3, x_2 = -\cos\theta, x_3 = \phi,$$

where r, θ, ϕ are the normal polar co-ordinates. These new variables are then polar co-ordinates but with a

---

[*] K. Schwarzschild, 1916, Sitzungsberichte der Königlich Preussischen Akademie der Wissenschaften zu Berlin, Phys.-Math. Klasse, 189(translation by S.Antoci & A.Loinger, arXiv:physics/9905030)

determinant of the transformation equal to one. Schwarzschild then proceeds to derive his solution and presents it in the form

$$ds^2 = (1 - \alpha/R)dt^2 - (1 - \alpha/R)^{-1}dR^2 - R^2(d\theta^2 + \sin^2\theta d\phi^2),$$

where $R = (r^3 + \alpha^3)^{1/3}$.

Hence, Schwarzschild's actual solution does contain a singularity when $R = \alpha$, but R is not the polar co-ordinate. It is clearly seen from above that, when $R = \alpha$, $r = 0$; that is, the singularity actually occurs at the origin of polar co-ordinates, as is usual. Therefore, according to Schwarzschild's own writing there is simply no singularity at $r = 2m$, to use the modern notation, and so the argument for general relativity predicting the existence of black holes cannot be justified by reference to the so-called Schwarzschild solution and it seems, as pointed out earlier, not a little ironic that non-rotating, uncharged black holes should be called Schwarzschild black holes.

It is not without both interest and relevance to note that, apart from Schwarzschild's own writings on this subject, other people of proven outstanding eminence in science had come to the same conclusion. As pointed out already, Einstein himself indicated his agreement with Schwarzschild's position in his 1939 article. The whole history of this emergence of the presently accepted so-called 'Schwarzschild solution' and, indeed, of the emergence of the presently accepted understanding of what a black hole actually is, is

The Schwarzschild Solution and Black Holes

covered in Stephen Crothers' article, *A Brief History of Black Holes*[*]. Most crucially in this history, he draws attention to the paper by Marcel Brillouin which dates from 1923[†]. As Crothers points out, Brillouin "obtained an exact solution by a valid transformation of Schwarzschild's original solution." He also showed rigorously that "the mathematical process, which later spawned the black hole, actually violates the geometry associated with the equation describing the static gravitational field for the point-mass." Further, he discussed the fact that the procedure adopted led to a non-static solution to what had been a static problem. This indicated quite conclusively a major contradiction. It showed that the solution was not one for the original problem! It might be remembered that the original problem which Schwarzschild set out to solve was to answer the question of what is the gravitational field associated with a spherically symmetric gravitating body, where the field is static (that is, unchanging in time) and the space-time outside the body is free of matter apart from a test particle of negligible mass. Schwarzschild did solve this problem. Others who followed manipulated his correct solution mathematically and ended up with 'solutions' which did not, and do not, satisfy the requirements of the original problem. Again as Crothers points out, this is very well documented. The names involved are of those well-known in science, both in the earlier years of the last century and now. Indeed, as is pointed out by Abrams[‡],

---

[*] S. J. Crothers, www.geocities.com/ptep_online/2006.html
[†] M. Brillouin, 1923, Le Journal de Physique et La Radium, 23, 43
[‡] L. S. Abrams, 1989, Can. J. Phys. 67, 919

Exploding A Myth

black holes, as discussed under the umbrella of general relativity, may be viewed quite reasonably as the legacy of an error by Hilbert. Abrams delves into the mathematics (a detailed discussion of which is not appropriate here) surrounding Hilbert's work and clearly identifies the error. His work is readily available in an easily accessible journal but is ignored, whether through ignorance or convenience only others can say. The entire sequence of events, or, in other words, the history of the unfolding of the present situation concerning black holes, is out in the open; it is simply not a set of facts hidden away in some dusty, forgotten archive. So how has the present situation evolved, where something which is known to be incorrect is widely accepted as scientific truth and, as such, attracts enormous public funding in one way or another? That has to be one of the biggest mysteries of modern science and the answer to it must be found, since it is so important for science as a whole, particularly in these days when funds are tight and so many projects are applying for funding.

These days, claims for the identification of black holes appear fairly regularly in the scientific literature. Quite often, the supposed existence of black holes - even that of so-called massive black holes - is invoked to explain some otherwise puzzling phenomenon. However, so far, on no occasion has the postulated object satisfied the requirement mentioned earlier that, for a black hole, the ratio of the body's mass to its radius - or more specifically in general relativistic

The Schwarzschild Solution and Black Holes

language, the radius of its event horizon - must be subject to the restriction

$$m/r \geq 6.7 \times 10^{26} \text{ kg/m} \quad *$$

Now it emerges that the mathematical singularity at the centre of the discussion simply did not appear in Schwarzschild's original solution of Einstein's equations. Obviously mathematics was used by Schwarzschild to find this solution, but it was used meticulously. It was noted carefully that, if a transformation of coordinates for which the determinant of the transformation does not equal unity is used, then the field equations themselves would not remain in an unaltered form. Hence, Schwarzschild adopted a transformation for which the value of the said determinant was one and went on to derive an exact - not approximate - solution to the equations. Also, Einstein himself proved that the singularity appearing in the popular form of the Schwarzschild solution has no physical significance. In all that Schwarzschild and Einstein did on this topic, the mathematics was a tool to help them achieve what they wanted. At no point was physical reality modified to fit a mathematical conclusion. This is the way things should be and provides an object lesson to many; - the mathematics is a tool and, as such, must be subservient to the physics.

Where then does that leave the modern notion of a black hole? Considerations such as those above,

---

* J. Dunning-Davies, 2004, Science, 305, 1238

undoubtedly raise major questions about the basis of much modern work. The idea of a body being so dense that its escape speed is greater than the speed of light remains not unreasonable but if the speed of light is a variable quantity - proportional, for example, to the square root of the background temperature, as suggested firstly by Thornhill[*], later by Moffatt[†] and, even more recently, by Albrecht and Magueijo[‡], as discussed in an earlier section here - many new and interesting questions arise.

As far as present day thinking goes, it is felt that, up to a certain limiting mass – the Chandrasekhar mass – a star will end up as a white dwarf, which is basically a degenerate electron gas. For a slightly larger star, it is felt it will end up as a neutron star – a degenerate neutron gas. At one time it was felt that that was as far as one could go and, for larger stars, the end point had to be a black hole, although no really sound scientific reason was advanced in support of this final conclusion. However, as yet, it is not known precisely how much mass stars eject before settling to their final form and it may be that stars eject enough mass during their lifetime to end up as either white dwarfs or neutron stars. On the other hand, it is thought now that neutrons are composed of quarks, and that quarks themselves may be composed of even smaller particles. This is, in fact, the current view. However, there are those who

---

[*] C.K. Thornhill, 1985, Speculations in Sci. & Tech., 8, 263
[†] J. Moffatt, 1993, Int.J.Mod.Phys.D., 2, 351; 1993,Found. Phys., 23, 411
[‡] A. Albrecht, & J. Magueijo, 1999, Phys. Rev. D., 59, 043516

The Schwarzschild Solution and Black Holes

feel that quarks are not actually physical particles but are a purely theoretical tool introduced to enable theoreticians to describe various properties of matter. This argument rages still but, for what follows, it will be tentatively assumed that quarks – or something very like them – do exist physically and are indeed particles obeying Fermi–Dirac type statistics. It then follows, as will be shown below, that stars larger than those which become neutron stars may eventually become quark stars, which are essentially composed of a gas of degenerate quarks; or sub-quark stars which are composed of a gas of degenerate sub-quarks. In all cases, these ideal models lead to end-point configurations possessing escape speeds less than the speed of light. Hence, the end points for the lives of stars of all masses could form a hierarchy of degenerate gases and the limit of such a hierarchy might be a black hole; that is, a body with an escape speed equal to, or greater than, the speed of light. However, that limiting case might, or might not, be attainable. In this area of stellar evolution, it is important to realise how little is truly known. Hence, it is vital to continue to keep *all* options open and to view *all* suggestions with a completely open mind, uninfluenced by the desires of the popular scientific press and the writers of science fiction.

The question of what happens to a star when it runs out of nuclear fuel will now be examined in a little more detail. It is found that the ultimate fate of any star depends on its mass. Low mass stars die in a continuous shedding of the star's outer layers, ejecting much of the

Exploding A Myth

mass into the inter-stellar medium in the form of planetary nebulae. Higher mass stars destroy themselves as supernovas. In both cases, the inter-stellar medium is enriched by a variety of heavy elements and the current view is that what remains is one of three types of exotic heavenly body; - white dwarfs, neutron stars or black holes.

Basically, stars go through a series of nuclear fusion reactions, starting with hydrogen and then helium burning. In a low mass star, less than about three solar masses, as much as twenty-five or even sixty percent of the star's mass is ejected in the form of a planetary nebula. In this case, helium is steadily converted into carbon in the core of the star, with hydrogen burning in a shell around it. This phase lasts for of the order of 10^8 years, leaving carbon but, in low mass stars, there is insufficient mass left for core temperatures to increase by gravitational contraction to a point where fusion of carbon is possible and so, thermonuclear reactions in the core stop. The outer envelope of the star is ejected into space in the form of a spherical shell of cooler, thinner matter called a planetary nebula. Once the nuclear burning ceases, there is no longer any outward pressure to resist the crushing force of gravity. As a result, the core is compressed to a size comparable with that of the earth and the density of matter rises to about 10^8 or 10^9 kg/m^3 - a teaspoonful of such white dwarf matter would have a mass of several tonnes! The star is prevented from further collapse, however, by so-called **degeneracy pressure**. Here a degenerate gas is one in which the particle concentration is so high that quantum

The Schwarzschild Solution and Black Holes

effects become important. If the particles are so-called fermions, which obey Fermi-Dirac statistical rules, the pressure is called the degeneracy pressure. This exceeds the normal thermal pressure because the particles obey the so-called Pauli Exclusion Principle. A consequence of this is that particles which are close together will possess different momenta and this difference in momentum is found to be inversely proportional to the distance between the actual particles, because of the uncertainty principle of quantum mechanics. As a result, in a high density gas, the relative momentum of the particles is extremely high and will not tend to zero as the temperature approaches the absolute zero of temperature. It is this type of mechanism which is thought to support both white dwarf and neutron stars against gravitational collapse.

A normal star may be modelled fairly accurately as an ideal ionised gas obeying the familiar equation of state

$$pV = nRT,$$

where p is the pressure, V the volume, n the number of moles of the randomly moving gas particles, R the universal gas constant, and T the absolute temperature. However, when the gas is compressed to the densities achieved in the interior of white dwarfs, this simple equation of state holds no longer. The particles cannot move about randomly but are squeezed together to the extent that electrons in neighbouring atoms tend to wish to overlap. The Pauli Exclusion Principle says that the electrons may not be compressed any closer together and, as a result, they exert a powerful outward pressure that opposes further contraction due to gravity. A gas in this state is called a degenerate gas and hence the notion of the core of a white dwarf being held together by the electron degeneracy pressure. It might be noted that this degeneracy pressure depends on the density of the gas **not** its temperature.

Stellar models show that the radius of a white dwarf is inversely proportional to the cube root of its mass; that is

$$RM^{1/3} = \text{const.}$$

This indicates that the more massive a white dwarf, the smaller it becomes. However, this is found to be true only to a certain upper limit known as the **Chandrasekhar mass.** This limiting value was derived in 1931 by the Indian astrophysicist Subrahmanyan Chandrasekhar, who showed that the maximum amount of mass a white dwarf may have and remain supported by electron degeneracy pressure is approximately 1.44 solar masses. It might be noted that many of these stars have been detected and none found so far violates this limit.

For stars whose mass exceeds this Chandrasekhar limit at the end of their thermonuclear burning phase, it is found that not even electron degeneracy pressure can prevent further collapse. The core temperature and density rise and gravitational contraction are so strong that electrons are pushed into protons so that neutrons are formed and neutrinos released in the process of inverse beta decay:

The Schwarzschild Solution and Black Holes

$$p^+ + e^- \rightarrow n + \nu$$

In a fraction of a second the core density becomes comparable with the density of an atomic nucleus; that is, 2×10^{17} kg/m^3. Such a final state is a **neutron star.** The upper limit to the mass for such an end-point in the life of a star is felt by some to be just two or three solar masses but other estimates go as high as four or even five solar masses. In this case, no precise agreed limit exists, although many do take it to be of the order of three solar masses. However, remember that this is the mass remaining at the end of the thermonuclear burning phase of the star's life and initially stars of mass greater than about 10 solar masses are those involved here. Most of the excess mass is removed during what is termed a supernova explosion; the neutron star is simply what remains after this spectacular event.

The centre of a neutron star is in the form of a superfluid gas composed of 80% neutrons, 10% electrons and the remainder protons. Note that a superfluid is one in which particles may flow over one another without friction. The neutron star is surrounded by a crust of iron covering a solid lattice of neutrons and neutron-rich nuclei. The superfluid neutron core is degenerate in the same sense as the electrons in a white dwarf and it is neutron degeneracy pressure which provides the outward force preventing any further collapse of the core and enabling the neutron star to become stable.

Further, neutron stars possess strong magnetic fields since the magnetic field of the original star is, at this

Exploding A Myth

stage, concentrated over a much smaller surface area, where the field strengths may be as high as 10^8T. Also, since the neutron star is so dense, it possesses a very high surface gravity - some 10^{11} times that of the Earth. Finally, for this reason, its surface is very smooth. Incidentally, the escape speed from the surface of a neutron star would be roughly 80% that of light.

It is felt by some that even this exotic situation is not sufficient to explain the deaths of stars of enormously high initial mass and that it should be possible - in principle at least - that the end point in the life of an extremely massive star should be a black hole. If such a situation did occur, the black hole would be of the type predicted by general relativity. However, before continuing the discussion in this direction, it might be noted that pulsars are basically rotating neutron stars. These exotic objects were first 'detected' accidentally by Jocelyn Bell who was working as a research student under the supervision of Anthony Hewish at a new radio telescope at Cambridge. Variable radio sources of extremely high and regular frequencies were detected and were called pulsars, being short for pulsating stars. These were found to be very compact galactic objects, much smaller and denser than white dwarfs. Hewish received the 1974 Nobel Prize for Physics for "his decisive role in the discovery of pulsars". However, pulsars are not pulsating stars, rather they are rotating neutron stars that formed in supernova explosions and it was Thomas Gold who finally put forward the full explanation for these objects discovered by Hewish and Bell.

The Schwarzschild Solution and Black Holes

It is of interest to note that the association of pulsars with supernovae, which was suggested by Hoyle as soon as pulsars were discovered, became accepted widely with the discovery of the Crab Pulsar with a period of only 0.033secs at the centre of the Crab Nebula which had been identified as the remnant of the 1054 supernova. The Vela Pulsar is yet another supernova remnant pulsar. This one has a period of 0.089secs and was discovered within the dispersed nebula of a supernova that occurred some 10,000 years ago.

Following the general idea of neutron stars came the question of what happens to stars of initial mass in excess of sixty (or some would say eighty) solar masses. It is felt currently, but not really known with any degree of certainty, that, after all the various processes are completed, the core remaining would possess a mass in excess of the limiting neutron star mass. When this question was posed first, nothing, other than the possibility of a black hole, seemed to remain as a possible explanation. Of course, it must still be just possible that, for supermassive stars, so much material is ejected in a variety of ways that the core left possesses a mass below the neutron star limit. However, if that is not the case, is a black hole the only possible outcome? Theoretically, this question must retain an open answer since, since that time, the idea of quarks and of all the so-called elementary particles being composed of combinations of the various types of quark

Exploding A Myth

has emerged. This surely opens up another route to be investigated.

Attention will be focussed now on some of the properties of dense matter. Very dense matter forms only a small fraction of the total mass of the Universe but has many interesting features. Generally it occurs in two situations:

One is at the end-point of stellar evolution when the nuclear energy sources are all used up and a new stellar equilibrium is established. This provides very dense bodies such as white dwarf and neutron stars as degenerate forms of ordinary matter. Dense matter in this form is distributed widely throughout a galaxy.

The second situation occurs specifically at the centres of many galaxies. A high concentration of mass is found here, often several powers of ten in terms of the solar mass. These regions are often confined within a radius of a light year (roughly 10^{16}m) or less and contrast strongly with other regions of a nebula, away from the nucleus, where such masses are spread over many light years.

An individual nucleon (proton/neutron) may be approximated as a sphere of mass 1.67×10^{-27}kg and radius 10^{-15}m. The mean density is then

$$\rho(n) = 3 \times 1.67 \times 10^{-27}/4\pi (10^{-15})^3 \approx 4 \times 10^{17} \text{kg/m}^3$$

This will be similar to the mean density of white dwarf and neutron stars and will be associated with radii in the range 10^7 to 10^4m. The mass of the entire visible Universe is estimated to be of the order of 10^{52}kg - 10^{11}

The Schwarzschild Solution and Black Holes

solar masses per mean galaxy and 10^{11} galaxies – and, with nuclear density, would be associated with a radius of some 10^{-5} light years. A region of dimension one light year at the centre of a galaxy would clearly not be of nuclear density. If so-called dark matter were included, it would increase the total mass by a little less than an order of magnitude and so increase the mass by a factor of approximately two and so, would not invalidate our various conclusions. Many galaxies seem to have a central region, of typical dimension one light year, containing a mass perhaps as great as 10^6 solar masses ($\approx 10^{36}$kg). This gives a mean density of 10^{-12}kg/m^3 for the region, a figure which is some 29 orders of magnitude less than nuclear density. The smoothed out density for the observable Universe is estimated to be of the order of 10^{-28}kg/m^3, so the central galactic region is very much denser than the smoothed mean. If nuclear densities are to be associated with the centre, it must be in a large collection of small units. In terms of white dwarfs and neutron stars, the region could contain an enormous number of components.

An interpretation based on General Relativity is that this central region contains a material singularity - a massive black hole. For a black hole of mass 10^{36}kg, the radius would have to be less than, or equal to, 10^9m or 10^{-7}light years. This has sufficient compression to explain the central region but the actual nature of the singularity still requires explanation.

Considering the title of this book, it seems in order, at this point, to introduce a small diversion which may serve as an illustration of the apparent power and influence of 'conventional wisdom'. It will illustrate also the different attitudes of the two journals said to possess the highest impact factors among scientific journals. In 2002, an article appeared in the British journal *Nature* (volume 419, page 694) announcing the discovery of a black hole at the centre of our own galaxy. It was noted immediately that the data presented did not satisfy the simple requirement for a black hole of the ratio of its mass to its radius being greater than, or equal to, 6.7×10^{26} kg/m. The journal was informed of this discrepancy but insisted that those complaining should first contact the authors of the original claim. This was done and those authors pointed out that they still felt the object they were examining was a black hole but the objection to the announcement that a black hole had definitely been identified at the centre of our galaxy was valid. *Nature*, however, would not publish the objection. An appeal registered with the Press Complaints Commission was rejected, even though they themselves had set a precedent in 1994 by insisting that *Nature* publish a letter pointing out elementary errors concerning entropy which had appeared in an editorial in that journal (*Nature* **356**, 103) in 1993. This critical letter eventually appeared in 1994 (*Nature* **368**, 284). In this latter case, *Nature* was, quite simply, ordered to publish. However, when it came to the later case concerning black holes, although the claim associated with the original article was shown quite clearly to be not proven and although it was not an editorial involved, the appeal was not upheld, thus seemingly violating a precedent. However, in 2004, a similar claim appeared in the American journal *Science.* In this case, the letter pointing out the problem was passed by the journal to the authors of the original article. They responded and *Science* subsequently published both criticism and response side by side on page 1238 of its issue of August 27th, 2004. This shows an enormous difference in attitude between these two dominant scientific publications. It is somewhat sad to note that one acted far more honourably than the other and, indeed, only one acted in a truly scientific manner. If open discussion is actively prevented, and that is basically what *Nature* was attempting in the case cited, science will find it increasingly difficult to progress. In a sense, it doesn't matter here who is right and who wrong. The fact is there was a genuine, potentially serious disagreement over a fundamental point of science. If such a disagreement is to be hushed up, it is science itself which will suffer in the end and, if such an action is allowed in one case, in how many others is it occurring also? It is possibly this final point which, in the grand scheme of things, is most important and worrying.

However, to return to the basic topic of black holes, the analysis of white dwarf and neutron stars is well-known. As mentioned earlier, both satisfy the mass/radius relation

$$RM^{1/3} = \text{constant},$$

where the constant equals $0.114(h^2/Gm_e m_p^{5/3})(Z/A)^{5/3}$ for a white dwarf and $0.114(h^2/Gm_p^{8/3})$ for a neutron star. The analysis leads to the radii of these two types of

Exploding A Myth

highly compressed star, if each is of about one solar mass, being 3×10^3 km and 1.5×10 km respectively.

Conditions for stronger compressions are not known with any degree of certainty but there is ample evidence to believe that hadronic matter is composed of quarks. Nucleons are regarded no longer as fundamental particles but, rather, are felt to be composed of quarks. These latter particles are unusual in two respects:

(i) they possess fractional charges ($\pm 1/3$ and $\pm 2/3$), the combinations of three quarks giving resultant charges of +1 for a proton and 0 for a neutron,

(ii) they require confinement together.

No free quark has been observed as yet, suggesting that they exist only within an environment of sufficiently high energy, although some arguments suggest that the quark coupling may be relaxed and the quark made "free" under appropriate physical conditions. In particular, there is a suggestion that the quark interactions become arbitrarily weak as the distance between the quark centres becomes smaller; this is often termed the ***principle of asymptotic freedom***. The full list of quark properties remains to be completed, although they are generally accepted to be Fermi particles. If this were not the case, they would not remain as independent entities within the nucleon particle, but might be expected to collect together as boson particles in a single state. Apparently, however, they can maintain their identity within their individual

The Schwarzschild Solution and Black Holes

'cells', even though they congregate together. To maintain a position within a cell suggests the ability to withstand external forces. Therefore, under appropriate conditions, there is no reason to expect the external gravitational force to be an exception.

It is not possible with the present state of knowledge to provide a full theoretical description of quark structures but significant conclusions based on energy considerations may be drawn. First, however, return to white dwarfs and neutron stars.

Quantum mechanics shows that a regime, including a specific particle, can be stable provided the particle is immersed in an energy field of magnitude less than its rest energy. Interest then centres on the rest energies of the electron, $\varepsilon(w)$, for white dwarfs and the neutron, $\varepsilon(n)$, for neutron stars. The relevant expressions follow from $\varepsilon = mc^2$, with $c = 2.998 \times 10^8$ m/s, and are

$$\varepsilon(w) = 8.18 \times 10^{-14} \text{J} \quad \text{and} \quad \varepsilon(n) \; 1.5 \times 10^{-10} \text{J}$$

Environments providing external energies of these magnitudes, or greater, will destabilise the particles. Here it is recognised that the stability of an individual hadron is possible because the forces of containment are normally such as to prohibit the occurrence of asymptotic freedom. However, presumably the stability of a hadron could be broken if sufficient external energy was provided.

In stellar applications, the external energy is provided by the gravitational energy of the star. The gravitational energy, E, of a spherical mass, M, of radius R is

Exploding A Myth

$$E \approx GM^2/R.$$

The body will contain $N = M/m_p$ particles, so the energy per particle, $\varepsilon(g)$, has the form

$$\varepsilon(g) \approx GM^2/RN = GMm_p/R.$$

The condition for stability is then of the form

$$GMm_p/R < \varepsilon$$

And this expression may be used to give a condition on the radius if the gravitational energy is not to become too large.

For a white dwarf, the condition for minimum radius occurs when

$$GMm_p/R < 8.18 \times 10^{-14} \Rightarrow R(w) > 3 \times 10^6 \text{m} = 3{,}000\text{km}$$

For a neutron star, the energy value given is that necessary for the neutron to dissolve. To retain the neutrons, it is necessary to assume the energy a little lower, say 10^{-11} J. Then, the required minimum radius is given by

$$GMm_p/R < 10^{-11} \Rightarrow R(n) > 2 \times 10^4 \text{m} = 20\text{km}.$$

These values are obtained, in each case, for a star of one solar mass. The values would be higher for stars of greater mass. If the maximum mass of a neutron star is taken to be three solar masses, the minimum radius would be about 60km. Then the neutron stars fill the mass range between the maximum value for a white dwarf and the value for true neutron evaporation.

The precise values obtained above are not significant but the orders of magnitude are because they indicate that arguments involving energy can provide useful results and a reliable basis for discussing sub-neutron structures. The hypothesis is made that the neutron will become destabilised when the external gravitational field exceeds the self-energy of the neutron. The minimum value for this to be possible is approximately 1.5×10^{-10}J. The quarks might be expected to be set free by a greater field and one might expect an energy of the order of 10^{-9}J to be involved; for a body of five solar masses, this would result in a radius of appreciably less than 10km.

It might be wondered if quarks themselves are the ultimate particles of matter and, indeed, it has been suggested, on the basis of preliminary observations only, that quarks themselves are composed of particles of mass 10^{-39}kg. If this were the case and such particles were, as is suspected, Fermi particles also, then a degenerate body made up of such particles could have a radius as small as 10^{-2}m. This would constitute an almost unimaginable degree of material compression. As an aside, it might be noted that 10^{-39}kg is the mass often attributed to æther particles by those who believe in the existence of an æther.

In postulating quark and sub-quark bodies, the level of singularity envisaged by General Relativity might appear to be being approached. However, there are essential differences between these various exotic bodies. The escape speeds for these exotic bodies

Exploding A Myth

introduced here remain less than that of light in vacuo. Hence, the particles always remain a part of the Universe. The escape speed for a black hole, on the other hand, is greater than that of light and so the black hole cannot have thermodynamic contact with the Universe. Also, the black hole, by its very nature, cannot radiate energy to the Universe directly but the dense objects considered here are in complete contact with the surrounding Universe; they are degenerate bodies with no internal heat sources and so will not radiate strongly. Again, the bodies considered here represent an equilibrium, both thermodynamical and mechanical, between gravity and known degeneracy forces. Therefore, they are part of the general hierarchy of degenerate bodies which lie in the range of weak compression of planetary satellites and the highly compressed density of neutron and sub-neutron stars. By contrast, the traditional black hole represents a non-physical entity, without either thermodynamical or mechanical equilibrium, possessing an entirely unknown internal structure. As a result, its physics is entirely unknown!

The one similarity between these two completely different descriptions of very dense astronomical matter (the sub-neutron star and the black hole) is that neither has, as yet, been definitely observed in practice, although there have been reports of the sighting of objects which appear to satisfy the requirements for a sub-neutron star – in the case in question, a quark star. The hall-mark of the black hole would appear to be its rapid rotation, as well as its small size and high mass.

The Schwarzschild Solution and Black Holes

The hall-mark of the sub-neutron star is probably its very small size and the very high mass of substantial numbers where they are contained in very restricted volumes. Whether physical conditions are possible for either object to exist is unknown. The centres of galaxies are often the designated regions for seeking black holes, but it is not completely clear why they should be there. The presence of sub-neutron stars could follow that of massive stars when the galaxy formed. It is likely these would have been so-called population III stars or something very close to them. The presently accepted theory of stellar structure suggests that the lifetime of a massive star is virtually independent of the mass and is a few million years. The evolution will have been passed through quickly and, if the star were very massive, a sub-neutron object could have been formed. This behaviour contrasts with that of low and medium mass stars where the lifetime is proportional to M^3. A star of solar mass will have a lifetime of about 10^{10} years and this is the mass to be expected of general stars in the galaxy. In this way, the central galactic region will differ from the outer regions.

In the discussion of the possible existence of black holes, attention so far has been restricted to a consideration of the fact that the relevant mathematical singularity does not actually feature in Schwarzschild's original solution of the Einstein equations and also on an examination of other possible end-points for the lives of stars dependent on the existence of sub-neutron particles. However, a further serious query hanging over the idea of a black hole, as an object seemingly derived from relativity, concerns the currently accepted theory for the thermodynamics of black holes.

In retrospect, it seems that it was inevitable that the analogy between an area theorem for black holes, published by Hawking in 1972, which asserted that, in any process involving black holes, the total area of the event horizon may only increase, and the established increase in entropy due to thermal interactions, was one that could not go unnoticed for long. If a connection was to be established, the question remaining was what function of the area was to be identified with the entropy of a black hole? The simplest choice compatible with Hawking's theorem is to set the black hole entropy proportional to the area of the event horizon itself. This choice was finally proposed by Bekenstein, also in 1972, and later – after apparently having some initial misgivings - endorsed by Hawking himself. It is important, in view of more recent developments, to note that actual order of events in the small piece of scientific history recorded in these few lines. The original proposal for the black hole entropy expression came from Bekenstein; it was accepted by Hawking at a later date. This is important in view of the total omission of Bekenstein's name from Hawking's book, *The Universe in a Nutshell* and also the adoption of the formula as Hawking's greatest achievement in the edition of the BBC Horizon programme broadcast in September of 2005. In the case of the television programme, the main discussion centred on problems associated with information being irretrievably lost in

The Schwarzschild Solution and Black Holes

black holes. However, this problem is another brought about by the adoption of the Bekenstein - Hawking expression for the entropy of a black hole. Thermodynamically, the expression implies possible violation of the Second Law of Thermodynamics. As far as information is concerned, the measure of information is the negative of the expression for the entropy. Hence, in this case, the information expression equivalent to the thermodynamic heat capacity will be a positive quantity, indicating information passing irretrievably into the black hole. Hence, two problems possessing a common source – the accepted expression for the entropy of a black hole!

Black holes are said to obey a 'no–hair' theorem. This states that black holes cannot be distinguished except for their mass, charge and angular momentum. In the simplest case of a Schwarzschild black hole, which is uncharged and non-rotating, the area of the event horizon is proportional to the so-called 'irreducible', or 'inextractable', part of the mass of the black hole. Actually, the entropy is postulated to have the form

$$S = kM^2/\sigma_m^2,$$

where M is the 'irreducible' mass of the black hole and $\sigma_m = (ch/2\pi G)^{1/2} = 2 \times 10^5$ gm is the Planck mass. The Planck mass is the quantity having units of mass that can be formed from a ratio of the fundamental constants and refers to the infant universe; that is, the universe when its age was approximately equal to the Planck time of $(Gh/2\pi c^5)^{1/2} = 2 \times 10^{-44}$ sec.

Exploding A Myth

Actual criticism of the established view has been minimal. However, it has been pointed out that, in conventional thermodynamics, the entropy is a first-order homogeneous function in all the extensive variables and this is not the case for this commonly accepted black hole entropy expression. (Here extensive variables, such as internal energy, volume and number of particles, are those which depend on the size of the particular system; all other variables, such as temperature and pressure, are termed intensive variables.) This might seem a somewhat trivial point to many people but it is, in fact, a feature which has several important consequences. In orthodox thermodynamics, one very useful formula is the so-called Gibbs-Duhem equation, which is a relation linking all the intensive variables of a system and shows that these variables are not all independent of one another. This formula has many important consequences and features in the derivation of many other formulae. However, the derivation of the Gibbs-Duhem relation itself depends critically on the extensive nature of the entropy of the system. Since the proposed black hole entropy expression is certainly not extensive in nature, it follows that there is no Gibbs-Duhem equation for such a system. Hence, formulae derived by using the Gibbs-Duhem relation must be excluded from use also when discussing such systems. It is possible that this is a technical point, which may be appreciated fully only by the theoretician but it is an important point which cannot be over-emphasised. The same argument may be employed when considering the derivation of the well-known Einstein – Boltzmann

The Schwarzschild Solution and Black Holes

formula for the probability of spontaneous fluctuations. This derivation holds no longer also. This follows because the Einstein formula implies that the entropy is an additive function; that is, if two systems are considered, the entropy of the combined system equals the sum of the entropies of the individual systems. Alternatively, this may be viewed as meaning that the joint probability of different events reduces to the product of the individual probabilities, implying statistical independence; in other words, the product of probability densities is tantamount to the sum of the entropies, which is Boltzmann's principle. Quite clearly, this is simply not possible for the present case because of the precise nature of Hawking's area theorem, from which it may be concluded that, if two black holes are combined, the entropy of a combined black hole is always greater than the sum of the entropies of the individual black holes, excluding the case where equality may hold. Hence, the Einstein – Boltzmann formula for a spontaneous fluctuation from equilibrium may not be used when considering thermodynamic black hole fluctuations. At the very least, this point has not been fully appreciated on a number of occasions and the said formula has been applied in a number of situations where its use is simply not permissible.

The fact that the sum of the areas before collision is not equal to the area after collision means that thermodynamic equilibrium may not be achieved. Consider two isolated systems at different temperatures. Suppose they are placed in thermal contact with one another but isolated from everything else. Eventually, in accordance with the zeroth law of thermodynamics, they will arrive at a common temperature. During this process, there will have been an increase in entropy. However, if the two separate systems had initially been at the same temperature, the entropy would not have increased. The above mentioned Bekenstein-Hawking expression for the entropy of a black hole is unable to cope with this particular, but very important, case, since, if M_1 and M_2 are the masses of the two black holes, then the mass after the collision is given by

$$(M_1 + M_2)^2 > M_1^2 + M_2^2.$$

Another important consequence of the presently accepted black hole entropy expression is that the heat capacity of the system is negative. Although such heat capacities are no strangers in astrophysics, inevitably they refer to one component, or phase, of a multicomponent, or multiphase, system. In reality a black hole must be an open system but it is always treated as a closed system. The mass could be written as the product $M = Nm$, where N is the number of 'particles' in the black hole having mass m, but, if N is not conserved, it would then be necessary to specify the second phase. Further, it has been shown possible for a negative heat capacity in a closed system to lead to a violation of the second law of thermodynamics and so, such heat capacities cannot be permissible. This point has been strengthened even more by work which indicates that it is the mathematical property of concavity of the entropy which embodies the essence of

The Schwarzschild Solution and Black Holes

the second law. Mathematically, the notion of concavity means that if a function f, say, depends on an independent variable x, say, then f is said to be concave with respect to x provided $\partial^2 f / \partial x^2 < 0$; that is, its second derivative with respect to x is negative. In other words, the graph of f drawn as a function of x would either curve upwards towards a maximum value before curving over and decreasing in value or be represented by part of a curve of this shape. In fact, this is merely a way of expressing mathematically the usual everyday linguistic meaning of the idea of concavity.

It might be argued that the second law, as popularly known, does not hold for such exotic objects as black holes. This is not a totally unreasonable point of view since the said law, although it might be said to have stood the test of time, is really a statement of fact based on worldly experience. For the hundred and fifty years or so since it was first proposed, people have sought to find violations of the second law of thermodynamics, just as they have striven to find violations of the first law. The reason for this preoccupation is the lure of 'getting something for nothing', while making massive inroads into the problem of solving the world's energy requirements. It goes without saying that, so far, all these efforts have been in vain. However, as pointed out by Planck if the units of time, length, and mass that may be constructed from the fundamental constants of nature "necessarily retain their significance for all times and for all cultures, including extraterrestrial and nonhuman ones, these 'natural units' would retain their natural significance as long as the laws of gravitation and the propagation of light in vacuum, and the two laws of thermodynamics retain their validity"[*]. Therefore, according to Planck, to question universality and the fundamental constants is tantamount to questioning the two laws of thermodynamics. Although it might be argued that it is not concavity, but rather the property of superadditivity (that is, the mass inequality shown above) that is the true stamp of entropy, it only needs one single exception to disprove this possibility. That exception is provided by black body radiation which possesses a subadditive entropy; where the property of subadditivity is exemplified by the mass inequality shown above but with the inequality sign reversed.

Since black body radiation has been mentioned, it seems worth considering, at this point, what happens when a black hole is bathed in black body radiation in a closed container? In the Bekenstein - Hawking entropy expression, the original dependence is on M, the so-called 'irreducible' mass. It is only via use of the relation

$$E = Mc^2$$

that the dependence of the entropy on the energy is established. Hence, for a so-called Schwarzschild black hole, the entropy is given by

$$S_{bh} = \frac{2\pi k G M^2}{hc} = \frac{2\pi k G E_{bh}^2}{hc^5},$$

---

[*] M. Planck, Ann. Der Phys. **4** (1901) 553

The Schwarzschild Solution and Black Holes

while that of black body radiation is given by

$$S_{bb} = \frac{4}{3} k \sigma^{1/4} E_{bb}^{3/4} V^{1/4},$$

where σ is the radiation constant. It needs a little imagination to achieve it but, given that, it might be possible to become convinced that the total entropy in the container is given by the sum of these two expressions; that is,

$$S = S_{bh} + S_{bb} = k \left\{ \frac{2\pi G E_{bh}^2}{hc^5} + \frac{4}{3} \sigma^{1/4} E_{bb}^{3/4} V^{1/4} \right\}.$$

The constancy of the total energy, $E = E_{bh} + E_{bb}$, means that $dE_{bh} = -dE_{bb}$ or, in other words, any changes in the black hole entropy must be exactly balanced by corresponding changes in the black body entropy. Again, the condition for thermal equilibrium demands that any change in the total entropy vanishes for arbitrary variations of energy. Hence,

$$1/T_{bh} = 1/T_{bb},$$

where, in an obvious notation, T_{bh} and T_{bb} represent the black hole and black body temperatures respectively.

Further, following earlier work, it might seem that

$$\frac{\partial^2 S_{bh}}{\partial E_{bh}^2} + \frac{\partial^2 S_{bb}}{\partial E_{bb}^2} < 0$$

is a condition for thermodynamic stability in a system comprising two bodies; in this case the two bodies being a black hole and black body radiation. From this last inequality, which expresses the concavity of the total entropy, there would result

$$E_{bb} < E_{bh}/4.$$

In Hawking's words, "in order for the configuration of a black hole and gravitons to maximise the probability, the volume, V, of the box must be sufficiently small that the energy E_{bb} of the black body gravitons is less than ¼ the mass of the black hole". However, thermodynamics can never place limits on the size of the volume or energy above which the system would be unstable. Thermodynamics is a 'black box' that provides no specific information about the system under consideration. An explicit physical model is necessary if actual numerical values are to be obtained.

In the situation just considered, a system composed of two parts – a black hole and black body radiation – was under examination. However, what precisely is a composite system? The notion of a composite system was introduced by Carathéodory, when he looked at the problems surrounding the foundations of thermodynamics at the beginning of the last century, in order to avoid considering nonequilibrium states. In fact, he compared two states of equilibrium, a more and a less constrained state of thermodynamic equilibrium that is achieved from the former by removing a restrictive partition between the two subsystems. Here the subsystems must necessarily be of the same type and not two different types, such as in the situation considered by Hawking[*]. It was claimed that "although

---

[*] S.W.Hawking, Phys. Rev.D **13** (1976) 191

The Schwarzschild Solution and Black Holes

the canonical ensemble did not work for black holes, one can still employ a microcanonical ensemble of a large number of similar insulated systems each with a given fixed energy E".

For anyone new to this field, it should be noted that all the material contained in the foregoing discussion is well documented[*] but, as is the case in so many walks of life, has been ignored by those who seem to regard their mission in life to be one of controlling what is, and is not, acceptable in science. So often, real scientific truth seems relegated to the sidelines and honest opinion ignored if it opposes 'conventional wisdom', even when that contrary view is supported by sound argument.

Before continuing with this discussion, it might be worthwhile to digress briefly to consider what is meant by an ensemble, and more specifically by the canonical and microcanonical ensembles mentioned. When considering actual physical systems, the physicist is faced with the major problem of how to deal with situations involving extremely large numbers. The answer is to resort to the methods of statistics and specifically to the so-called ensemble method of Gibbs. Gibbs' method is based on several fundamental postulates or axioms, in much the same way as Euclidean Geometry is based on a number of basic axioms. The validity of the suggested approach rests on the agreement between experimentally derived results

---

[*] J. Dunning-Davies, 1996, Concise Thermodynamics, (Albion Publishing, Chichester), B. H. Lavenda, 1991, Statistical Thermodynamics, (John Wiley & Son, Chichester),B.H.Lavenda, 1995, Extreme Value Statistics, (Albion Publishing, Chichester), and references cited in these books

Exploding A Myth

and those deduced directly from the theory. So far, the theory had stood up to all the tests. In this approach, an ensemble is simply a collection of a large number of systems, each made as a replica on a thermodynamic level, of the actual thermodynamic system under investigation, - always remembering that thermodynamics is a macroscopic theory and reveals nothing of the microscopic structure of a system. As an example, the system of interest may consist of a volume V containing N particles all of the same type, and be immersed in a large heat bath at temperature T. These three variables prove sufficient to determine the thermodynamic state of the system. For such a case, the ensemble would consist of a large number of copies of the system, all identical from the thermodynamic viewpoint. However, they would not all be identical on the level where the individual particles come in for consideration. The three thermodynamic variables mentioned would prove totally inadequate to specify the detailed microscopic state of the system which could contain an extremely large number of particles. For the example mentioned, the pressure is not actually specified and indeed its value could be different in the different copies of the original system. The theoretician would average these values over all the systems of the collection, or ensemble, leading to a value for the so-called ensemble average of the pressure. A similar procedure could be adopted to find the ensemble average value for any mechanical variable which may have different values in the various systems of the ensemble.

The Schwarzschild Solution and Black Holes

This technique was, as stated earlier, developed by Gibbs and has proved extremely successful and useful. It might be noted that the technique might almost be thought the physicists' answer to mathematical statistics. It is a matter of some regret to realise the two topics have progressed, over the years, virtually independently of one another when, in reality, the physical theory might be thought the physical manifestation of mathematical statistics. Statistical mechanics, as the theory due to Gibbs is now called, could have benefited tremendously if it had been developed in parallel with mathematical statistics; much duplication of effort could have been avoided and many results, deduced as apparently new in statistical mechanics, could have simply been adopted from mathematical statistics or one of its branches. A typical example where much duplication of effort could have been avoided concerns the modern area of the examination of so-called non-extensive systems where the vast majority of the 'new' results are in fact rehashes of results know to information scientists for years; but more of that later.

Now to return to the subject of ensembles in physics; the actual ensemble discussed as an example was one for which the values of the volume, number of particles and temperature were given. Such an ensemble is the one known as a canonical ensemble. If, instead of the temperature, T, the value of the internal energy, E, had been given, or fixed, the resulting ensemble would have been a microcanonical ensemble. These, of course, are the two ensembles to which reference is made in the

Exploding A Myth

quote from Hawking. Obviously, any ensemble is a purely theoretical device used to enable the theoretician to describe a particular physical situation and, hopefully, to make predictions about it. As usual, everything is based on a model of reality and so, the accuracy of the predictions provides some measure of the value of that model.

It should be noted that energy is normally assumed to be conserved: only the form in which it appears and the carriers of it change. Further, 'energy does not transform into anything, it is only different particles that transform into one another'. Hence, the rest energy of a black hole may not be transformed into energy of black body radiation or vice versa and so, what one loses, the other does not necessarily gain.

Hence, several outstanding questions remain unanswered relating to the modern idea of a black hole. Physically, it is relatively easy to imagine Michell's idea of what is essentially a dark body and shouldn't properly be referred to as a black hole. It is simply a body whose density is such that its escape speed is greater than the relevant speed of light. Here the word 'relevant' is included merely to allow for the possibility of the speed of light varying in some way. Such a body, if it existed, would simply possess a very strong gravitational field. Of course, bodies would be able to leave its surface but, depending on their initial velocity of projection, they would only rise so far into the atmosphere before being dragged back by gravity. – in much the same way as a ball is dragged back to the surface of the earth after being thrown upwards into the

The Schwarzschild Solution and Black Holes

air. It follows that such a body would be clearly visible to observers, provided they were close enough to the body. It is simply the case that nothing projected from the surface of such a body, unless it was able to achieve a speed of projection greater than the speed of light, would be able to totally escape the gravitational influence of that body. However, the more modern notion of a black hole, while often referred to as the end point in the life of a star of large mass and therefore linking it directly with a genuine physical structure, seems nevertheless to be occasioned by an almost desperate need to attribute a physical interpretation to a mathematical singularity. In this case the singularity apparently appears in the Schwarzschild solution of Einstein's equations and so it is something which seems to emerge from Einstein's General Theory of Relativity. As discussed earlier, the major immediate objection concerns the fact that the offending singularity did not actually appear in Schwarzschild's original solution and is simply a property of the coordinate system chosen so often for use. Further, a careful examination of the present thermodynamic theory associated with black holes reveals more and more inconsistencies. All this would, of course, have to be viewed somewhat differently if a black hole had been identified beyond all reasonable doubt. So far, many candidates have emerged but none has so far passed the relatively simple test of satisfying the inequality, relating to the ratio of the mass to the radius, first derived via Newtonian mechanics by Michell in 1784. The debate will, no doubt, continue to rage but, in future, let such debate be

Exploding A Myth

scientifically driven, not obliterated by the power of 'conventional wisdom' and its influential supporters.

As far as the actual general topic of black holes is concerned, what is the present state of knowledge as to their actual existence? A black hole is an object obviously well-loved by science fiction writers and, indeed, by devotees of science fiction. As such, it would be lovely if they existed but, from their point of view, all that is really needed is lack of knowledge indicating that they definitely do not exist. From the scientific viewpoint, much time, effort and money has been invested in both the theoretical investigation of black holes and their properties and in the search for them in the cosmos. Failure to find them might be thought by some to be totally unacceptable, especially since so much money has been spent on the search already. That is possibly the major problem facing science as far as black holes are concerned – the seeming existence of overwhelming outside pressure to establish the definite existence of these objects. That is why it is so important to view every claim of finding such an object very critically. Maybe such objects will be identified. If so, it will be vital to consider in detail precisely what they are. Such objects would of necessity possess a mass and radius which satisfies

$$M/R \geq 6.7 \times 10^{26} \text{kg/m}.$$

However, although such an object would possess an escape speed greater than or equal to the speed of light, would it be the type of object envisaged using Newtonian ideas or would the ideas of relativity come in to play? This is a vitally important issue because the Newtonian ideas are well established, what might be termed the relativistic view is based almost totally on, what to many, is an incorrect solution of the Einstein field equations attributed erroneously to Schwarzschild, as pointed out earlier. If anything, it is this latter point which is the major one to be addressed in this context. It would be useful also if, in the not too distant future, astrophysicists could devise a theory covering all stages in the life of a star, regardless of its initial mass, – from birth to death. This would involve knowing, or at least having some idea, of how much mass a star divests itself during its life before entering its final stages. After all, as far as the present state of knowledge is concerned, one possibility must be that any star, however massive, will eventually divest itself of sufficient mass to ensure that it doesn't end up as a black hole. It does seem that, as yet, there is no truly compelling reason for discounting this possibility. Obviously, if black holes are definitely found to be at the centres of galaxies, that still leaves open the questions of what they really are and how they originally formed. If black holes are found elsewhere, as has been claimed recently by scientists using NASA's X-ray Timing Explorer[*], the basic theory of what constitutes a black hole will still need to be reviewed because, as has been amply demonstrated, the model based on the erroneously named Schwarzschild solution cannot be deemed acceptable. It is interesting to note that the wording of this latest announcement

---

[*] www.physorg.com./news9693.html

Exploding A Myth

appears to imply absolute acceptance of the existence of black holes in our universe as well as an implied dependence on numerical data as a justification for the claims made. It should be remembered that numerical models are always totally dependent on the data originally fed into the chosen computer programme. This final point raises questions in all instances where there is dependence on computer generated information, whether in the field of black holes or elsewhere. In these days when computers intrude so much into peoples' lives – whether professionally or otherwise – and where computer error is the favourite excuse advanced to explain mistakes, it is vital to remember that the computers themselves do not generate errors; errors are due to the operators and specifically due to the incorrect structuring of a computer programme or the incorrect input of data. It will be interesting to see how these recent claims concerning the identification of black holes stand up to detailed, unbiased scrutiny.

The immediate future beckons with interest, but it is vital that any issues raised in the context of black holes are viewed with totally open minds; strict adherence to the presently accepted 'conventional wisdom' could delay progress for many years in much the same way as Lord Rayleigh claimed progress in theoretical physics had been delayed by some anonymous referees' rejection of the paper by Waterston on the kinetic theory of gasses. This is, of course, referring again to the case of Waterston as discussed in the earlier chapter concerned with Einstein's theories of relativity and it is absolutely clear

The Schwarzschild Solution and Black Holes

why the case of Waterston is, indeed, relevant to the present discussion. In so many situations – the Big Bang and black holes, to mention but two – 'conventional wisdom' combined with an enormously powerful status quo has exerted an unrelenting pressure on the ways in which scientific research should proceed and possibly on the actual topics that were, and indeed are, investigated. The case of Waterston, particularly after it was highlighted by Brush, should have served as a salutary warning to the scientific establishment to be extremely wary of such courses of action. It is often claimed that people produce their best work at an early age. If that is so, such people would have no opportunity to establish themselves in the scientific community, as Lord Rayleigh suggests very strongly they should, before attempting to publicise what could easily be revolutionary ideas. It also highlights the effects of burying Waterston's ideas; Lord Rayleigh points out that, after 1860, all reference on the subject would have been made to Maxwell and so, the effect of ignoring Waterston's work amounts to retarding scientific advances in that particular field for approximately fifteen years. Lord Rayleigh also comments that the referee concerned was "one of the best qualified authorities of his day" and he does not wish to be judgemental as far as that person is concerned. It may be worthwhile reflecting that, in science as in all other areas of human activity, it is not always wrong to be judgemental; people should not fear this label of being judgemental as many are made to do in our modern, so-called enlightened society. However, the grave warning for science as a whole is there for all that hath eyes to see and ears to hear and, possibly more importantly, minds open to receive!

Chapter Five

Hadronic Mechanics

Introduction

For roughly one hundred years now, science has almost complacently drifted along in a publicly assumed belief that Einstein's relativity theories, together with quantum mechanics, offered the true means of solving all the remaining theoretical problems of science. Put in this bald fashion, many would throw up their hands in horror. However, that is the situation the world of science has been facing for some time. For years, undergraduates have been told of the complacent attitude existing at the end of the nineteenth century when, the story goes, many eminent scientists believed all the theoretical tools necessary to solve all the world's scientific problems were known; it was simply a matter of time before all the answers were found. This sort of sentiment was, and is, often used as a prelude to introducing the theories of relativity and quantum mechanics in university undergraduate courses. The story also served to ridicule the scientific establishment in place at the end of that century. It might be thought that a lesson would have been learnt from this story, but no. At the end of the twentieth century, eminent

Exploding A Myth

scientists once again vociferously proclaimed the same position as that so falsely claimed at the end of the previous century. It seems that eminent scientists, like most other people, can be tempted very easily into making rather foolish claims in order to gain a little – in fact, a very little – short-term, high profile publicity, or should it be more properly be called notoriety? This is possibly a lesson the non-scientific public should learn. However eminent a scientist may be and in whatever field of science, he is still human and, as such, is prone to human frailties and mistakes like everyone else. It is often claimed the public, through the media, places individuals – be they sportsmen, politicians, philanthropists or scientists – on pedestals, only to destroy them if they err. There may be some truth in this assertion but surely, therefore, it is sensible not to allow oneself to be placed on such a pedestal in the first place? The rewards may be great, but the fall is so much greater!

In the present case, however, what of these claims concerning the theories of relativity and quantum mechanics? As has been seen already, there are grave qualms over the theories of relativity harboured by many people, but what of quantum mechanics? There have been worries expressed over some points in quantum mechanics almost from the very beginning of the subject. Frequently, these have revolved around the role of the observer and over whether or not quantum mechanics is an objective theory. One man who has considered these points at length is Karl Popper, probably one of the best known philosophers of science.

Although he has written on the topics at length, his book *Quantum Theory and the Schism in Physics*[*] proves an excellent source of his views. He expresses the view that the observer, or, as he prefers to call him, the experimentalist, plays exactly the same role in quantum mechanics as he does in classical physics; that is, he is there to test the theory. This, of course, is totally contrary to the so-called Copenhagen Interpretation, which provides the normally accepted position. This alternative view basically claims that "objective reality has evaporated" and "quantum mechanics does not represent particles, but rather our knowledge, our observations, or our consciousness, of particles".[†] As Popper points out, there have been a great many very eminent physicists who, over the years, have switched allegiance from the pro-Copenhagen camp. He cites among these Louis de Broglie and his former pupil Jean-Pierre Vigier, Alfred Landé and, in some ways most importantly, David Bohm. Bohm, himself an acknowledged and deeply respected thinker, wrote a book on quantum theory, which was published in 1951[‡], in which he presented the Copenhagen point of view in minute detail. Later, apparently under Einstein's influence, he arrived at a theory[§] "whose logical consistency proved the falsity of the constantly repeated dogma that the quantum theory is 'complete' in the sense that it must prove incompatible with any

---

[*] K. R. Popper, 1982, Quantum Theory and the Schism in Physics (Hutchinson, London)
[†] W. Heisenberg, 1958, Daedalus, 87, 95
[‡] D. Bohm, 1951, Quantum Theory, (Prentice-Hall Inc., New Jersey)
[§] D. Bohm, 1966, Reviews of Modern Physics, 38, 453

more detailed theory'"[*]. It was this very question of whether or not quantum mechanics is 'complete' which formed the basis of the intellectual struggle between Einstein and Bohr. Einstein said 'No'; Bohr claimed 'Yes'. The whole problem is discussed in great detail by Popper and, for those interested in this important topic, there can be no better reference than the book by Popper mentioned already.

However, where does Popper fit into anything to do with Hadronic Mechanics? Quite simply, the answer lies in the fact that it was in his 1982 book[†] that he, Karl Popper, drew attention to the thoughts and ideas of Ruggero Santilli. In the 'Introductory Comments' to his book, Popper reflects on, amongst other things, Chadwick's neutron. He notes that it could be viewed and indeed was interpreted originally as being composed of a proton and an electron. However, again as he notes, orthodox quantum mechanics offered no viable explanation for such a composition. Hence, in time, it became accepted as a new particle. Popper then notes that, around his (Popper's) time of writing, Santilli had produced an article in which the "first structure model of the neutron" was being revived by "resolving the technical difficulties which had led, historically, to the abandonment of the model"[‡]. It is noted that Santilli felt the difficulties were all associated with the assumption that quantum mechanics

---

[*] K. R. Popper, 1982, Quantum Theory and the Schism in Physics (Hutchinson, London)
[†] Ibid
[‡] R. M. Santilli, 1981, Foundations of Physics, 11, 383

applied within the neutron and disappeared when a generalised mechanics is used. Later, at the end of section IV of his 'Introductory Comments', Popper makes the following assertion:

"I should like to say that he (Santilli) – one who belongs to a new generation - seems to me to move on a different path. Far be it from me to belittle the giants who founded quantum mechanics under the leadership of Planck, Einstein, Bohr, Born, Heisenberg, de Broglie, Schrödinger, and Dirac. Santilli too makes it very clear how greatly he appreciates the work of these men. But in his approach he distinguishes the region of the arena of incontrovertible applicability of quantum mechanics (he calls it atomic mechanics) from *nuclear mechanics* and *hadronics*, and his most fascinating arguments in support of the view that quantum mechanics should not, without new tests, be regarded as valid in nuclear and hadronic mechanics, seem to me to augur a return to sanity: to that realism and objectivism for which Einstein stood, and which had been abandoned by those two very great physicists, Heisenberg and Bohr".

Obviously, these comments of Popper will not be too well-received by some but, at the very least, they provide much food for thought and, considering his own well-deserved reputation, should convince people to assess Santilli's contributions with open minds.

As stated above, in more recent times, one man who has worried about the extent of the claims for these theories, both relativity and quantum mechanics, is Ruggero Santilli. He has devoted his life to studying them and attempting to extend the theories to cover situations to which they were not, in their usually accepted forms, truly applicable. The fact that they are, at the very least, not applicable in certain cases is something which is hidden from the public and from most students and Santilli's investigations have placed him squarely in opposition to the 'godfathers' of 'conventional wisdom'. All this has put him at a grave disadvantage in the scientific world and there seems little doubt that without the unswerving support of his wife, he would not have survived. Incidentally, it is worth noting, at this point, that this last statement is so true of so many who have opposed 'conventional wisdom'. Their wives have offered unstinting support ungrudgingly. For this, these courageous women – for that is precisely what they are - should be saluted and thanked by the entire scientific community, for their quiet moral support has to be recognised as a major factor in helping their husbands continue with their work in the face of so much hostile, scientifically unwarranted - indeed bigoted - opposition.

However, returning to the whole story surrounding Ruggero Santilli, as already noted, he has dedicated his life to examining the bases of relativity and quantum mechanics, feeling both theories to be incomplete. His investigations have led, in recent years, to possibilities for new clean energies and it is this which is now so important to consider, especially at this time when the world is so troubled by the depletion of energy stocks and worries about environmental effects of the energy sources presently being utilised so widely. This whole

Hadronic Mechanics

problem of future energy supplies is probably far more serious than usually imagined. Present demand is increasing but, when countries such as China, the Indian sub-continent and those of Africa come on line fully and require as much energy as the countries of the present west, that demand will escalate enormously. Given the present state of orthodox fundamental knowledge, the only realistic solution to this problem is presented by nuclear power. To many, this is not an acceptable option. Alternatives such as solar power, wind power, geothermal energy, wave energy, and others are all put forward but, in truth, these in total would come nowhere near satisfying the probable future demands for energy. No; as has been pointed out on several occasions[*], the only realistic answer at the world's disposal at present is nuclear power. However, nuclear power is felt to pose two major problems and both are concerned with safety. The safety of the actual power stations is, not unreasonably, a tremendous worry for many. This is accentuated by incidents such as the Three Mile Island problem in the U.S.A. and, more recently, the disaster at Chernobyl. However, it is only the latter case that proved a true disaster; the first was fundamentally contained by the safety systems in place. There is little doubt that, provided adequate funds are made available, nuclear power plants can be made extremely safe, although, as with all man-made

---

[*] G. H. A. Cole, 1996, in Entropy and Entropy Generation, ed. J.S.Shiner; Kluwer Acad, Pub., Netherlands.

V. Castellano, R. F. Evans and J. Dunning- Davies, Nuclear Power and the World's Energy Requirements, arXiv:physics 0406046

Exploding A Myth

structures, no-one can guarantee complete safety of anything and, whether those in authority like to admit it or not, genuine accidents will, and do, occur. Therefore, there can be no room for complacency but, if a sensible number of safety measures are incorporated into the plant, nuclear power stations should be safe. The disposal of nuclear waste, however, is another matter, as has been highlighted by all the problems being faced in the U.S.A. over its proposed storage facility in Nevada. This brings the story back to Santilli for another outcome of his work has been the emergence of a possibility for the safe disposal of nuclear waste in-house; by which is meant, the safe disposal of the waste without any need for transportation. The idea is still only at the theoretical stage and, as Santilli has been requesting for some time now, requires the performance of about three experiments to see if the theory actually works in practice. Such experiments would not be cheap to perform but, considering the enormous sums spent on some elementary particle work, the cost would not be too great and, if successful, the ensuing benefit for mankind would truly be out of all proportion to that cost!

Most will ask at this point why these experiments haven't been performed. This is a difficult, if not impossible, question to answer, but it may be noted that, on the one hand, the theory behind all this does not conform to 'conventional wisdom' and does, in fact, raise questions about the range of validity (at least) of the widely accepted theories of relativity and quantum mechanics, while, on the other hand, the theory has led already to the production of the new clean fuel, 'magnegas'! Hence, although the theory may be abstruse, may contain elements which some feel unacceptable, and may conflict with 'conventional wisdom', nevertheless something concrete has been produced which can be, and has been, used. The theory definitely appears to have had a readily identifiable success already. On the other hand, enormous profits are being made by people in the business of disposing of nuclear waste using the current somewhat crude and unsatisfactory methods. So the question arises as to whether, in some sense, 'conventional wisdom' and 'big business' have combined to prevent the performance of these experiments which, if successful, could have such a dramatic effect on both.

While the details of magnegas and its production are readily available via the internet (at www.magnegas.com or www.i-b-r.org) and may be read about in Santilli's book the *Foundations of Hadronic Chemistry*[*], it is worth noting that it was in 1998 that Santilli first built a so-called hadronic reactor of molecular type – something also known as a PlasmaArcFlow reactor. Such reactors make use of a submerged DC electric arc to achieve the recycling of nonradioactive liquid waste into a clean combustible gas called 'magnegas'. The process involved also produces heat, which may be used via exchangers, and some solid precipitates. These reactors provide an ideal means of disposing of most kinds of liquid waste –

---

[*] R. M. Santilli, 2001, Foundations of Hadronic Chemistry (Kluwer Academic Publishers, Dordrecht)

sewage, oil waste, other contaminated liquids and so on, - but may be used to process fresh or salt water also if necessary. In the above- mentioned book, Santilli comments that the best liquid for use in these reactors is crude oil, which may be processed into an extremely clean combustible gas at a fraction of the cost of normal refinery processing. However, the use of crude oil would hardly be beneficial in the present circumstances.

As is described in detail by Santilli[*], the said reactors operate by liquids flowing through a submerged DC arc with at least one consumable carbon electrode. The arc decomposes both the liquid molecules and the carbon electrode into a plasma at approximately $3,500°K$. This plasma is composed predominantly of hydrogen, oxygen and carbon atoms. The plasma is moved away from the arc as soon as it is formed and the reactor controls the recombination into 'magnegas', which bubbles to the surface where it is collected. Due to the known affinity of carbon and oxygen, oxygen may be removed from the plasma which results in combustible carbon monoxide. The removal, in turn, of this carbon monoxide, as soon as it is formed, then prevents its oxidation into carbon dioxide and so reduces the carbon dioxide content of the gas dramatically. The hydrogen essentially recombines into hydrogen molecules, although there are other products also.

---

[*] Ibid

Hadronic Mechanics

The use of an underwater arc is, of course, nothing new but, in other apparatus, the resulting carbon dioxide content of the emerging product is unacceptable environmentally. This is one of the bigger points in favour of this new technology. Again, the large glow normally created in underwater arcs is due to the recombination, following separation, of hydrogen and oxygen into water. This, of course, helps account for the low efficiency of the said underwater arcs. The new reactors, however, display a dramatic increase in efficiency due, at least in part, to the removal of hydrogen and oxygen from the arc immediately following their creation, thus preventing recombination into water. This hugely increased efficiency is a major plus for these new reactors and results in the production of a combustible gas at a price which is genuinely competitive with the cost of fossil fuels. When this overall cost is considered, it must be remembered that it will be arrived at after any income derived from the recycling of liquid waste and the utilisation of the heat produced has been taken into account.

'Magnegas' is largely unknown in many parts of the world and so, having introduced it as above, it is worth realising that it has been subject to extensive testing. The results are impressive! A Ferrari 308 GTSi and two Honda Civics have been converted to use 'magnegas'. One of these vehicles has been the subject of the above-mentioned testing. It has been found that 'magnegas' exhaust surpasses all the usual safety requirements without the use of a catalytic converter; emits no harmful carbon monoxide, carcinogenic or

Exploding A Myth

other toxic substances in the exhaust; reduces carbon dioxide emission due to petrol combustion by roughly 40%; and actually emits some breathable oxygen. This final fact is highly unusual since most fuels act to deplete the oxygen in the atmosphere; this one enhances it! However, not only is this final fact unusual, it is possibly highly important since, if the world continues with its present activities, what effect will oxygen depletion of the atmosphere have eventually? With all the talk of the dangerous environmental effects of present energy policies, oxygen depletion of the atmosphere is one rarely, if ever, mentioned. A further point of possible interest to motorists with a passion for performance cars, is that use of 'magnegas' as fuel doesn't seem to affect performance adversely, - at least not by much. In fact, a 'magnegas' fuelled Ferrari was privately raced successfully against conventionally fuelled Ferraris.

To conclude this introduction and justification for this chapter, it seems worthwhile to include a direct quote which appears in Santilli's book *Foundations of Hadronic Chemistry*, page 283. The quote is attributed to Mr. John Stanton, President of EarthFirst Technologies Inc., who states that

"the new technology of PlasmaArcFlow Reactors is evolutionary, rather than revolutionary, because conceived to be primarily beneficial for crude oil, piston engines and hydrogen industries".

The interest in this claim is the use of the word 'evolutionary'. This is the way in which the work is perceived by an industrialist. How would, or indeed do, academics view it? It would surprise a great many people if the overall academic judgement failed to err on the side of revolutionary and actually moved to condemn Santilli's work for that very reason.

Hadronic Mechanics.

The book of Santilli's discussed in the introduction was published in 2001 and was produced to provide a possible explanation for a number of problems which had persisted for many years in the general area of quantum chemistry. After a century of research, despite a great many successes, a number of basic issues remained unresolved by orthodox quantum chemistry. Among these were:

(i) the lack of an exact representation of molecular data when derived from first principles, with deviations of the theory from experimental data on binding energies of the order of 2%;

(ii) the inability to permit accurate thermochemical calculations, since 2% is missing in the representation of the binding energies, corresponding to about fifty times the typical energy releases of thermochemical reactions, such as that in the formation of the water molecule;

(iii) the absence of an attractive valence force sufficiently strong to explain the strength of molecular bonds existing in nature;

(iv) the inability to restrict valence bonds to electron pairs only, thus essentially implying the prediction of molecules with arbitrary numbers of constituents;

(v) the incorrect prediction that all molecules are paramagnetic.

Obviously, the origins of Santilli's work go back much further and the applications are already much wider than is implied by the words 'hadronic chemistry'. The list of books, apart from all other publications, is impressive but contains mention of virtually all his contributions to various areas of science. However, whether he is considering a problem in astrophysics or biology, as he himself says, he approaches it as a mathematical physicist. Also, he took as his starting point a seemingly unshakeable belief in the idea that science, in general, doesn't admit complete and final theories, and could not progress without the introduction of some new mathematics. One immediate example illustrating this is provided by Newtonian mechanics, which had been so successful for so long, finding itself being regarded as a special limiting case of relativistic mechanics towards the beginning of the last century. Also, Einstein's general theory of relativity brought to the fore in the world of physics new mathematical methods. This new mathematics involved tensors and was reliant on earlier work by such as Riemann, Ricci and Bianchi. Hence, the huge change in physics at the beginning of the twentieth century was accompanied by new mathematics being introduced and used in physics and a well-established theory clearly being seen to be approximate and not final. Accordingly, Santilli turned his attention to producing new mathematics in order to deal with these new problems. To do this, he turned to the work of Marius Sophus Lie for some of his inspiration. After much intellectual effort, Santilli proposed so-called hadronic mechanics which is basically an image of quantum mechanics formulated via several completely new forms of mathematics, termed by him iso-, geno-, and hyper-mathematics, with so-called isoduals for antimatter. The corresponding iso-, geno-, and hyper-mechanics are then found to represent single-valued reversible, single-valued irreversible, and multi-valued irreversible systems respectively. Fundamentally, hadronic mechanics preserves all the usual laws and principles of orthodox quantum mechanics but represents what might be termed a completion of that subject, as seemingly required by the well-known argument of Einstein, Podolsky and Rosen (*The Physical Review*, **47**, 1935, 777). It is strongly suspected by many that Santilli's hadronic mechanics genuinely achieves this objective. However, the whole truth will be known only after the wider scientific community has examined the veritable mountain of material with an open mind. Incidentally, the names for these three new branches of mathematics/mechanics were constructed for the following reasons: firstly the 'iso' prefix, being short for isotopic which comes from the Greek and is meant to indicate the property of axiom-preserving for the new theory; secondly, the 'geno' prefix comes from genotopic which again follows from its Greek meaning which suggests an axiom-inducing property of that new theory; and finally, the term hyperstructural basically arose from ideas of multivalued functions. Further, iso-mechanics is fundamentally a non-unitary theory but is reversible; geno-mechanics preserves this property of non-unitarity but introduces ideas of irreversibility; hyper-mathematics goes even further and, while preserving non-unitarity and irreversibility, introduces multi-valuedness which increases the number of degrees of freedom open to the investigator and thus permits the study of far more complicated structures than was allowed previously.

It is not intended to discuss the precise details of any of these new forms of mathematics or, indeed, mechanics here but, suffice it to say, that a major difference between the forms of mathematics proposed by Santilli and the form with which everyone is so familiar, is that Santilli proposed using something other than the usual 'one' as the unit for his mathematics. For example, in the simplest form used for investigating anti-matter, the unit is -1, instead of 1. A very simple introduction to the use of this particular case is furnished by the examination of the associated thermodynamics as discussed in *Thermodynamics of Antimatter via Santilli's Isodualities* (*Found. Phys. Lett*, 1999, **12**, 593-599). In other forms, the structure proves more complicated. This immediately indicates that the new theories should be capable of discussing more complex systems of nature than was possible for classical and quantum mechanics since those theories only had real and complex numbers at their disposal. This limit placed by orthodox mathematics on mechanics, both classical and quantum, might be felt to be responsible, at least in part, for such theories having to make such suppositions as all particles being point-like.

At this point, it might be remembered that mathematics has long been termed the language of physics, but, with the principles of physics extending into so many regions of science these days, it might be termed the language of science more appropriately. That being the case, it is not too surprising if major changes, or even extensions, have to be introduced into our mathematical preconceptions when it comes to dealing with totally new situations.

To say that some of these situations are totally new might be thought something of an understatement, given some of the areas to which the new mathematics has been applied successfully. However, before considering that, it might prove beneficial to consider some reservations of current knowledge expressed by leading scientists of earlier years. Santilli himself admits to a lasting impression being left on him by several of these. In his book *Nuclear Physics* (University of Chicago Press, 1950), Fermi states on page 111 that "there are doubts as to whether the usual concepts of geometry hold for such small regions of space (those of nuclear forces)". This is, by itself, an extremely powerful statement by one of the leading scientific figures of his age but is it well-known, do people pay it due attention?

Exploding A Myth

The answer to both those questions is probably 'No'. It is of further interest that the dedication of Santilli's book *Elements of Hadronic Mechanics*, Vol. 1, (Naukova Dumka Pub., Kiev, 1995), is to the memory of Enrico Fermi "*because of his inspiring doubts on the exact validity of quantum mechanics for the nuclear structure.*" Santilli also alludes to a statement included in Blatt and Weisskopf's book *Theoretical Nuclear Physics* (John Wiley, 1963) in which they speculate on page 31 on the possibility "that the intrinsic magnetism of a nucleon is different when it is in close proximity to another nucleon". In fact, this statement acted as a major spur to Santilli who claims to have produced a complete theory of total nuclear magnetic moments via his so-called hadronic generalisation of quantum mechanics. Whether or not he has achieved this is for the scientific community as a whole to decide but, until his work is read with open minds and properly digested, no final verdict can be sensibly announced. This indicates, once again, the urgent need for a totally open-minded examination of Santilli's work. A third, possibly rather obvious, source of inspiration was provided by the very well-known article by Einstein, Podolsky and Rosen which appeared in the journal *The Physical Review* (volume 47, page 777) in 1935, This article voiced concerns about quantum mechanics and it is worth realising that, until the day he died, Einstein continued to harbour real doubts concerning the lack of deterministic character of quantum mechanics. This again raises the question of how the scientific community, in general, regards Einstein. To many in the general public it is probably felt that he is still revered as the greatest scientist of the twentieth century. If that is the case though, it seems surprising that so many of his views and beliefs seem to be misrepresented. By referring back to his writings of the early years of the last century, it soon becomes apparent that here was a man who wrote very precisely and with great clarity. A good example of this is provided by his writings on Brownian Motion, now collected into a small book, *Investigations on the Theory of the Brownian Movement* (Dover, 1956). The writing in this small volume could well serve as an object lesson to all writers of science. Nevertheless, as commented on earlier when discussing black holes, there are several occasions where his views are kept well hidden, and have been kept so hidden for many years. It is not without significance to note that it is some of the truly 'big' names of twentieth century science who were voicing these qualms about the total validity of quantum mechanics over many years of the last century. These were also people who, it is well-known, were highly articulate. There was, and is, no reason to doubt what they were saying or about the grave doubts they were harbouring. In many ways the scenario is a repeat of that facing relativity in the earlier years, at least, of the last century. It is a sobering thought that, by this time, some may be wondering how science has managed to progress as far as it has, and with so much success. The added thought, however, has to be how much farther mankind might have progressed if unhampered by 'conventional wisdom' and all its attendant trimmings.

Exploding A Myth

Another spur to Santilli's investigations was provided by the realisation that most of contemporary physics is concerned with the examination of systems subject to conservative fields of force; that is, subject to forces which are derivable from potentials. A good everyday example is provided by the gravitational field which so markedly affects our everyday lives. This is the example with which so many are familiar from school and which forms the basis for the introduction to the ideas of kinetic and, more importantly in the present context, potential energies. If the motion of an object held at arm's length before being released to fall to the floor is considered, it is seen to gather speed until it strikes the floor. At the instant before it actually strikes the floor, it is at zero distance above the floor but is moving at its highest speed during the entire motion. At that point, all its energy is said to be kinetic; that is, all its energy is due to its motion. However, at the initial moment of release, the object is not moving and so, has no kinetic energy. All its energy is due to its height, its position, above the floor. This energy is said to be potential energy; it is the energy which the object possesses because of its position and which gives it the potential for movement. This potential energy is purely due to the presence of the gravitational field, whose action pulls the object towards the centre of the earth or, in this case, towards the floor. This gravitational field is one of those force fields said to be conservative because potentials are associated with them. All the basic mechanics taught in schools and universities is done so under this restriction. Only rarely are situations for which there is no potential energy discussed. In a way this is not unreasonable since so much that affects us directly is governed by conservative fields of force. Newton's mechanics incorporating conservative fields of force are found to describe accurately both very small systems and very large systems. These days, problems of astronomy are considered at a variety of levels by everyone in his own home via well-established television programmes such as *The Sky at Night* and, since planetary motion is thought to be governed by a conservative field of force, this serves to reinforce the notion of such fields being all important, so that the possibility of non-conservative fields is often forgotten or even ignored. However, when the original writings of such as Lagrange and Hamilton on the analytical approach to mechanics are examined, no such restriction is apparent. This is certainly not clear in the vast majority of, if not all, undergraduate courses on Analytical Mechanics, as the whole area is commonly called. Restriction to conservative fields of force occurs at a very early stage. Of course, in fairness, to the undergraduate this does not seem at all unreasonable. Whether it be the mathematician or the physicist, the majority of actual situations met will be concerned with conservative fields of force. To a large extent, the same excuse for absence of consideration of more general situations from undergraduate lectures is valid but, in reality, attention to this restriction should be drawn. In mathematics lectures, no-one would contemplate drawing back from making *all* restrictions placed by a theorem crystal clear. This must be the correct approach, even though, in most cases, those restrictions will not affect the practising physicist. As has been pointed out on numerous occasions, when dealing with problems of the physical world, the physicist is regularly warned to be careful that mathematical restrictions on the use or applicability of a result may be coming into play by the physics of the situation. A perfect example to illustrate precisely what is meant by this is provided by the phenomenon of phase transitions. In the case of water, for example, it is patently obvious that something very unusual is happening when ice changes into liquid water and when that liquid water turns into steam. In both cases, it is observed that, at particular temperatures, as heat is added to the system, the structure of the substance changes but the temperature remains fixed. This is contrary to what is normally believed to happen when heat is added to a system. Hence, at these two temperatures of $0°C$ and $100°C$ for water, something very unusual is happening physically. This should alert the physicist to be wary; to be very wary of what is happening physically but, in some ways more importantly, to be wary of whether or not mathematical expressions remain valid. This is another good example of the use of mathematics as the language of physics; here the applicability of mathematics is ruled by the physics of the situation – not the other way around!

To return specifically to Santilli's contributions, it is remarkable to note to how many different outstanding problems he has turned his attention with this new approach and, apparently, with so much success. As mentioned, one of his earliest worries concerned the range of applicability of quantum mechanics. Having noted the comments and concerns of some truly notable scientists of the early part of the last century, he devised so-called Hadronic Mechanics and succeeded in explaining a wide variety of otherwise unexplainable phenomena. These are catalogued in detail in his book *Foundations of Hadronic Chemistry* but it is worth noting, and speculating on, some of them here; one in particular being particularly relevant to something which has preceded it, but more of that example later. As noted on page 35 of his book, explaining the experimental data on the Bose-Einstein correlation in proton - anti-proton annihilation at both high and low energy provided experimental verification of hadronic mechanics in particle physics. Such experimental data may be represented by traditional quantum mechanics only after the introduction of arbitrary parameters which seem to have no physical origin. However, hadronic mechanics is easily able to explain things because it proves capable of dealing with the off-diagonal terms appearing in expectation values. This latter property is not allowed in orthodox quantum mechanics because, for a quantity to be observable, its expectation value must be diagonal in form. This, of course, introduces mathematical terms into the discussion which, ideally, should be avoided but, suffice it to say, that the phenomenon may not be explained by orthodox quantum mechanics because it is too restricted as a theory. Another experimental verification, in the sense of the previous example, has been provided by the ability of the new theory to explain data concerning the anomalous behaviour of the mean-life of the kaon with energy. This has been examined successfully over various energy ranges and is important because, as with the example of the Bose-Einstein correlation, it establishes the existence of effects in the interior of kaons which are nonlinear, non-local and, most importantly, non-potential (that is, non-conservative).

As Santilli has stated quite categorically on several occasions but, possibly most clearly at the beginning of section 3 of his article in the *Journal of New Energy* (1999, **4**, page 106), he has always thought of physical particles as being particles which may be defined rigorously in our spacetime. He points out that hadronic mechanics was conceived and developed in order to identify the constituents of all unstable hadrons with genuine physical particles. Has he succeeded? Time will tell, but the positive evidence is there for all to see and is mounting. As has been seen already, any discussion of this topic inevitably seems to introduce mathematical ideas and notation at some point. Again as stated already, this is unfortunate but doesn't detract from an appreciation of the picture emerging and might serve as a spur for professionals to investigate the detail further in order to reach a truly informed opinion of the work.

From the point of view of physics, it seems that Santilli obtained inspiration from early ideas of Rutherford. It was in 1920 (*Proc. Roy. Soc. A*, 1920, **97**, 374) that Rutherford postulated the existence of a new particle, which was, in essence a 'compressed hydrogen atom'; that is, it was composed of an electron compressed entirely within the proton. This he called a neutron. Presumably Rutherford thought that, when a hydrogen atom is compressed, for example, in the core of a star, the high pressures involved could result in it being reduced in size to that of a proton, with an electrically neutral particle emerging finally. Twelve years later, Chadwick (*Proc. Roy. Soc. A*, 1932, **136**, 692) established the existence of the neutron experimentally. However, Rutherford's original conception of this particle was dismissed by many of the founders of quantum mechanics for a variety of seemingly good reasons at the time: - the model would require a positive binding energy; both constituents possess spin ½ and so, the resulting particle would not be permitted to have spin ½ by normal quantum mechanics; orthodox quantum mechanics would also not allow the correct magnetic moment to follow in this model. Hence, the rejection of Rutherford's model of a neutron and this heralded a change in the direction of physics' research. Up to that time, physics had been based on the notion that the constituents of so-called bound states have to be capable of being isolated and identified in laboratories. The rejection of Rutherford's conception appears to have altered this view. This then was the spur for Santilli and, having devised the new mathematics referred to earlier, he first succeeded in producing a consistent model of the meson, π^o, as a bound state of an electron and a positron. This model is not possible in conventional quantum mechanics for a number of reasons, one of which concerns binding energy. Quantum bound states possess negative binding energies and this implies a total mass less than the sum of the constituent masses. For a π^0 meson, this would imply a rest energy appreciably less than its actual rest energy of 135Mev. This problem, as are all others, is resolved by hadronic mechanics or, at least, that is the claim with all the evidence clearly available for examination by those with a mind so to do. The model Santilli proposes does, in fact, explain all the characteristics of the said particle – zero spin, electrically neutral, null magnetic moment, a rest energy of 135Mev, a mean-life of approximately 10^{-16}sec., a charge radius of about 1fm (that is, 10^{-15}m), decay according to

$$\pi^0 \rightarrow e^+ + e^-,$$

- and this model of the smallest of hadrons has now been extended successfully to all mesons. Further, although the theory does not view quarks as actual physical particles, but rather as mathematical objects with a composite structure, this new model for hadrons does prove compatible with the current quark theories, always assuming that quarks have a composite structure. For those interested, further details of this model may be found in a variety of publications but especially in volume **4** of the *Journal of New Energy*, as mentioned earlier. In fact this reference is a veritable goldmine of information on this general topic of hadronic mechanics and its consequences both for physics itself and probably for mankind as a whole through its consideration of the possibilities offered by the theory for alternative new clean energies.

However, what could conceivably turn out to be Santilli's most important achievement was his success in using the new hadronic mechanics to resurrect the Rutherford model for the structure of the neutron successfully. This model recognises a neutron as being composed of a bound state of a proton and an electron at a distance of 1fm; that is, at a distance of 10^{-15} m. As mentioned earlier, such a model is prohibited by conventional quantum mechanics, so, if Santilli's ideas are valid, what are the consequences for physics? The answer is, quite simply, enormous! The abandonment of the original approach to the structure of physical particles will have had a profound and far-reaching effect on research in the area of particle physics obviously. However, it is the possible ecological implications which are staggering and of so much direct relevance to absolutely everyone. The orthodox approach has conceivably prevented the study of the neutron as a major source of clean energy and actually seems to have obstructed the study of new forms of clean nuclear energy. These are now being studied via hadronic mechanics, as is the associated problem of the safe disposal of the nuclear waste presently causing so much trouble.

The main characteristics of the neutron, such as its having a rest energy of 939.6Mev, a mean-life of 916 secs., spin ½, and a charge radius of 0.8×10^{-13} cm., were all explained in a model of the neutron devised by Santilli using hadronic mechanics in 1990 (*Hadronic J.* **13**, 513). This was a non-relativistic treatment, but a relativistic treatment soon followed and appeared in 1993 (*JJINR Comm.* E4-93-352). The crucial point about this is that the model was precisely that proposed by Rutherford so many years earlier. Using hadronic mechanics, Santilli was able to derive all the properties of the neutron when it was viewed as being composed of an electron totally compressed inside a proton. This model, remember, had been abandoned because this structure was inexplicable using orthodox quantum mechanics. However, the fact that the Rutherford model may be explained using this new technique cannot, in itself, be regarded as justification for the new hadronic approach. The real justification is provided by the fact that there appears to be experimental verification of the structure in that experimental verification of the synthesis of neutrons from protons and electrons seems to have been achieved in the 1980's by a group in Brazil under C.Borghi, although the results were published only in 1993 (*J.Nucl.Phys.* (Russian) **56**,147). Although this is exciting, it is by no means conclusive evidence and that is precisely why caution is exercised when reporting and discussing this development. However, the possible ramifications are so important that it is vital for this experiment to be repeated several times so that a genuine conclusion may be reached which may be accepted by all in the scientific community.

The ramifications alluded to concern the possibility of utilising these new theoretical ideas to produce new clean energies for mankind. This again is a topic to which Santilli has devoted much time and energy over the years. Basically, many of these new energies are characterised by processes in the interior of hadrons, rather than in nuclei or atoms. It might be noted that energy is required if unstable hadrons are to be synthesised from physical particles; in the case of the neutron, 0.80Mev is required to synthesise it from protons and electrons. However, as Santilli points out (*Journal of New Energy*, 1999, **4)**, "once created, unstable hadrons become a large reservoir of energy, which is released in their decay". Some of these proposed new energies, therefore, are produced by using mechanisms capable of stimulating the decay of unstable hadrons, or by simply using the energy produced in their natural decay. In this article, he goes on to describe the way in which energy could conceivably be produced via stimulated neutron decay. He also draws attention to the quantity of energy involved, pointing out that the electron emitted in neutron decay would possess energy roughly 100,000 times more than that of electrons hitting a computer screen. Again, it is noted that this mechanism is possible only if the neutron is composed of the physical particles, the proton and the electron. The main ideas behind the proposal are that the neutron does actually decay spontaneously. Also, its mean-life is not fixed but depends on local conditions; for example, if it's a constituent of some unstable nuclei, the mean-life is a few seconds; in a vacuum, it's more of the order of fifteen minutes; in other unstable nuclei, it's even longer; and in natural, light, stable nuclei, it's infinite. However, the neutron itself is naturally unstable and so it is felt it should be possible to stimulate its decay and hence control its mean-life. The actual proposal suggests testing this possibility through the use of photons with the resonating frequency of 1.204Mev,

plus the additional threshold energy required to satisfy conservation requirements of
$$\gamma + n \rightarrow p^+ + e^- + \nu.$$
Here the figure of 1.204Mev for the resonating frequency is another consequence of the hadronic model of the neutron adopted. It has been found, by studying nuclei, that most nuclei do not permit reactions such as that represented by the above equation due to violation of conservation laws. However, some do and it is these which offer the possibility of a new form of usable energy, termed by Santilli *hadronic energy*. In his book, Santilli chooses, as a representative example, Molybdenum ($_{42}Mo^{100}$) but also draws attention to the fact that other natural, light elements, such zinc ($_{30}Zn^{70}$), possess the required prerequisites. Most of this is still in need of experimental verification. It seems that, if successful, these tests would offer a prize too valuable to be ignored. It is to be hoped, therefore, that the necessary experiments will be performed in the very near future, so that existing doubts may be cleared up, one way or the other, finally.

A further important reason for having the predictions of hadronic mechanics fully and openly tested is provided by the rapid accumulation of highly radioactive nuclear waste around the world. This is proving a major problem for many countries. The U.S.A. has been seen to have a major problem of disposal and also to have an additional problem posed by those opposed to the current method for attempting to achieve that disposal. Britain, on the other hand, while facing problems concerning disposal of its own nuclear waste, faces additional protests from those opposed to its business of helping in the disposal of nuclear waste from other countries. In both instances, and in others, people are extremely worried by the perceived threat posed by the actual disposal method as well as that posed by the transportation of that waste across country. All of these worries have been exacerbated by the rapidly growing terrorism threat facing so much of the world. There can be no doubt that a great many people, some with scientific knowledge, some without such knowledge, harbour genuine worries. There can be no doubt also that those worries, and indeed fears, are not unjustified. The above discussion surrounding the composition of the neutron obviously offers the possibility of a resolution of the difficulties and concerns. These essentially reborn ideas concerning the structure of the neutron, if valid, offer the possibility of recycling nuclear waste by way of stimulating its decay in such a way as to reduce the extremely long lifetimes to hours or, at worst, days. It is envisaged that this could be achieved by the use of relatively light equipment and that the nuclear power plants could achieve this within their own boundaries, thus eliminating all transportation of these highly dangerous materials. If the idea works, although jobs in the industry presently formed around the disposal of nuclear waste would vanish, many new jobs in a much safer nuclear waste disposal industry would appear. The new industry might be expected to grow for the development, production and sale of the new equipment, since it would be a vital requirement for nuclear power plants throughout the world.

Exploding A Myth

The basic idea revolves around the fact that the nuclei concerned are large and naturally unstable. One idea is to expose the highly radioactive nuclear waste to an intense, coherent flow of photons with the required resonating frequency. It is felt that this may be achieved via a synchrotron of about three metres diameter; - a size which could be accommodated in nuclear power plants. A typical example is provided by uranium ($_{92}U^{238}$) which has a life-time of the order of 10^9 years. A double stimulated transmutation of this element could change it into Plutonium ($_{94}Pu^{238}$). Again, this is an unstable quantity and has harmful emissions as well, but its life-time is a mere 86 days and it could well be retained under suitable shields for that period of time. It may be superfluous to draw extra attention to this point, but it is worth noting the different life-times involved here – 86 days as against 10^9 years! The phenomenal advantage of this stimulated transmutation is immediately evident. Will it work? The theory certainly suggests that it should, but only experimentation will give the actual answer to that question. Possibly the bigger, more relevant, question to ask at this time is whether or not the scientific community and national governments are prepared to finance the experiments necessary to test this thesis?

At this moment in time, it is worth realising that the cost of carrying out the proposed experiments would probably be of the order of a few hundred thousand pounds. This sounds a lot of money, and indeed it is. However, an experiment to detect neutralinos – those particles predicted by theory as candidates for so-called 'dark matter' which seems so important to preserve the currently accepted standard model in cosmology - has been running for sixteen years with no success so far. Nevertheless, it has been announced recently that those running this experiment are installing yet another new detector at the cost of one and a half million pounds! It has also been announced recently that, in America, a new extremely powerful super-computer has been used to create a three-dimensional model of two colliding black holes. Since this is purely computer experiment, it must be noted from the very outset that any results obtained will be totally dependent on the original input model and information. Both these factors will be completely dependent on present day knowledge and, possibly more importantly, theories. Hence, both will be influenced heavily by 'conventional wisdom'. Nevertheless, the results from this computer experiment are being heralded as very exciting and it is proposed to use this information to restart another sequence of very expensive experiments to seek evidence of such collisions, including yet another search for gravitational waves. This latter search is again, incidentally, another extremely expensive series of experiments which has continued for a great many years with, as yet, absolutely no success. This second proposed venture has not been costed as yet but will undoubtedly eat up millions of pounds of scientific research money. Fundamentally, no-one interested in science should be opposed to either of these two possible areas of research. Both will add, either positively or negatively, to human knowledge and, as such, are important. However, even if

Exploding A Myth

successful, neither will produce any immediate major benefit for mankind. If a few hundred thousand pounds were to be spent checking out Santilli's theories, the worst that could happen would be negative results; in which case a few hundred thousand pounds would have been wasted, but yet again, knowledge would have been gained. Negative knowledge may be, but knowledge nevertheless. If successful though, mankind's energy worries would recede into the background, at least for the immediate future, and nuclear power would become a so much safer option. Also, with the problem of the disposal of nuclear waste dealt with so that the genuine worries of so many would be assuaged.

However, the scientific establishment tends to regard orthodox quantum mechanics as a sacrosanct part of 'conventional wisdom', so it must be thought doubtful that it will sanction work which directly challenges that 'foundation stone of modern science'. The positions of national governments are far more difficult to assess. They will consult scientific advisers who will be members of the scientific establishment, so the line of their advice is probably predictable. They will be under pressure from a wide variety of areas of 'big business' but, no doubt, the most vociferous will be those wreaking profits from the present highly questionable methods of nuclear waste disposal. They will also, though, be under pressure from members of their electorates. If news of this possibility of there being a truly safe, in-house method of disposing of nuclear waste did become fully public, then it is probably this final factor that would weigh most strongly with national governments since, at the end of the day when all the political manœuvering and gesturing has been discarded, it is the thought of votes at the next election which would end up being of paramount importance. Can the possibility of the existence of such a prize really be ignored any longer?

The success in describing the above mentioned model for the neutron using this new hadronic mechanics opened the way to view afresh models for other systems, in particular the deuteron. Here an unresolved problem had lain around for years; that was the inability of conventional quantum mechanics to explain the value of one for the spin of the deuteron. The deuteron was felt to be composed of two particles, each having spin a half and the basic axioms of quantum mechanics would imply, therefore, a spin value of zero for the ground state of such a system. The new hadronic mechanics clears up this problem also. Following on from the reduction of the neutron to an hadronic bound state of a proton and an electron, the deuteron is viewed as a three-body situation comprising two protons and one electron – or, more accurately in Santilli's language, two iso-protons and one iso-electron. This model is able to represent accurately all the characteristics of the deuteron, including its spin. This success led Santilli to extend the notion to all nuclei. The result was to produce a new hadronic structure model of nuclei in terms of combinations of iso-protons and iso-electrons, which reduces to the usual model involving protons and electrons as a first approximation. This all seems at first sight to be merely another huge amount of almost unintelligible theory which will have little or no effect as far as the ordinary person is concerned. Amazingly, that is not the case. If this theory does turn out to be correct, the implications for society are immense because it could result in a number of new forms of clean energy for mankind's use; forms which are not possible with the old proton – neutron model. It does appear, therefore, that this is an area worthy of further open-minded investigation simply because the possible prize at the end is so attractive and, indeed, necessary considering the massive environmental problems and energy demands facing our world at the moment.

Further Applications.

So far, the applications discussed have been associated with elementary particles. It has been seen that, from this area alone, many benefits for mankind as a whole could accrue, if the predictions of the theory prove both accurate and achievable in practice. However, although a major factor in inspiring the researches which have led to these was the concern about energy resources, other fields may benefit from the development of these new mathematical techniques also. An unresolved problem facing astrophysics is one mentioned earlier, and that is the assertion by Arp that some quasars are physically linked with galaxies which appear to possess completely different redshifts. This assertion is based on, and supported by, a substantial body of observational evidence. Arp himself has offered an explanation, which revolves around the actual meaning, or interpretation, of the observed redshifts for objects. As mentioned earlier, he suggests that the redshift possesses two components and only one of these is the so-called Doppler shift; the other being an intrinsic component. The present position is, of course, to discount the interpretation of Arp's observations that the quasars and galaxies are linked physically and to continue to interpret the different redshift values as meaning that the quasars and associated galaxies are at totally different distances from the earth and are moving at totally different speeds relative to us. This, of course, is to interpret it simply as a Doppler type shift, is in line with 'conventional wisdom' and agrees with the accepted Einsteinian treatment of cosmological redshift. In 1991, using his new mathematics, Santilli suggested another explanation, (see *Isotopic Generalization of Galilei and Einstein's Relativities*, vols. I & II, Hadronic Press, 1991). His suggestion amounted to the difference being accounted for by a slowing down of the speed of light within the chromospheres of the quasars. It should be realised that these chromospheres are thought to be extremely large and the suggested effect is very similar to the slowing down of the speed of light within our own atmosphere. The result of this suggested slowing down would be for the light to leave the quasars – or more correctly, the quasar chromospheres – already redshifted. As far as the individual stars of the galaxy are concerned, they are effectively isolated in space and are thought to have dramatically smaller chromospheres. Hence, for the stars of the galaxy, the effect alluded to here will not exist. The end result is that, for physically connected quasars and galaxies having exactly the same expansion speed, the light from each will reach us here on earth with dramatically different redshift values. The reason advanced for the new theory being more suitable for explaining this effect is that traditional theory assumes everything both isotropic and homogeneous. It is thought, however, that chromospheres are both anisotropic and inhomogeneous. Hence, the need for utilising Santilli's iso-mathematics and related results to explain these observational results originally highlighted by Arp as discussed earlier. A further consequence is, of course, that redshift is not necessarily a measure of the expansion of the universe. This thought is not one to be accepted too readily by current adherents to 'conventional wisdom'. However, in the *Journal of New Energy*, volume **4**, evidence supporting this claim is presented clearly on page 103, where it is noted also that another verification offered within astrophysics for this new theory is provided by the quantitative – numerical representation of the internal red and blue shift of quasars. Basically, it seems that the cosmological redshift for each individual quasar is not constant but actually depends on the frequency of the light with an internal redshift for the infrared part of the spectrum and an internal blue shift for the ultraviolet part. These mean an increase and a decrease respectively of the cosmological redshift for these parts of the spectrum and are, of course, totally incompatible with special relativity since they imply different speeds of light for different frequencies in the interiors of quasar chromospheres. This behaviour is, however, predicted exactly by Santilli's modified theory. The studies associated with this topic also indicate that one contribution towards the red-sky viewed on occasions at both sunrise and sunset is isotropic in origin. The idea is that the anisotropic, inhomogeneous structure of the earth's atmosphere provides an additional contribution to the redshift at sunset since, then, the earth's rotation simulates motion away from the source. It is thought, therefore, that the larger redshift observed at sunset, as opposed to sunrise, is due to the rotation of the earth.

Again, when biological structures are investigated, it soon becomes clear that one of the biggest differences between those and the more usual physical systems is their non-conservative character. This latter thought is becoming more and more important in the present day as biology becomes more and more dependent on mathematics and theoretical physics in its development in some directions. At present, the biggest area where this occurs is possibly in the theory behind evolution where thermodynamics is playing an increasingly important role. Indeed, the Second Law of Thermodynamics really is appearing to look as if it may be one of those laws of nature whose influence pervades most, if not all, areas of science and even beyond. However, as far as Santilli's work is concerned, the power and range of applicability of his new mathematics is apparent when the problem of the growth of sea-shells is considered[*]. As he himself points out, it emerges that Euclidean geometry, with which

---

[*] R. M. Santilli, 1996, Isotopic, Genotopic and Hyperstructural Methods in Theoretical Biology, (Naukova Dumka Pub., Kiev)

Exploding A Myth

most are so familiar, is insufficient for a consistent representation of the actual growth of sea-shells; the possible shapes of sea-shells are represented perfectly well by Euclidean geometry with no need for any extension into broader theories, but the generalised methods, introduced by Santilli, become vitally important when a detailed examination of the growth in time of these sea-shells is required. One major problem is that the growth of sea-shells is definitely non-conservative and also irreversible. However, the problem was eventually solved by Illert and Santilli[*] using the new iso-euclidean geometry as developed by Santilli. The use of the alternative geno-euclidean geometry might have proved more appropriate in some ways since it might allow for a deeper axiomatisation of irreversibility. Obviously, studies such as those alluded to here are in the early stages of applying this new mathematical structure to biological problems. It remains to be seen how widely this new mathematics will be used but, initially, the results of applying it to a wide range of problems are good and so it is to be hoped that mere 'conventional wisdom' will not hinder its future use in even more fields.

---

[*] C. Illert and R. M. Santilli, 1995, Foundations of Theoretical Conchology, (Hadronic Press, Florida)

Chapter Six

'Conventional Wisdom': Some Modern Case Studies

The common theme running through all the earlier chapters has been the influence of what has come to be called 'conventional wisdom' on science. There is no doubt that factors, separate from true science, have affected the progress of most, if not all, branches of science for many years; the case of Waterston, dating from the middle of the nineteenth century, is a classic case in point. However, that particular case is well in the past; lessons should have been learned from it. It now appears that no such lessons have been learnt. Factors other than purely scientific ones still appear to be exerting tremendous influences on progress in a wide variety of fields. These factors would appear to range from simple personal jealousy to the protection of individual interests, both scientific and possibly even those of big business. It may be idealistic, if not naïve, to expect that science should remain pure and stay unaffected by such factors but, if science is to progress satisfactorily, these external factors must be held at bay.

Although the earlier chapters, except possibly for the first chapter, have concentrated on the general position

Exploding A Myth

of certain issues in science, the history of some aspects of a few quite specific examples will now be discussed in detail to illustrate just how these influences may be, and are, brought to bear on situations. It might be pointed out from the outset that one of the most successful techniques adopted is one used in so many walks of life – silence! It is surprising how often people simply ignore criticism unless forced to respond by being, quite literally, caught out in public. It is a ridiculously simple technique but one which works so effectively on so many occasions in so many totally different circumstances. When you are faced with an awkward problem or criticism, you simply remain quiet on the matter in the almost sure and certain hope that, in time, the problem will simply go away! How simple! How dangerously effective! How devastatingly destructive when used ruthlessly in the field of science!

1. Black Hole Entropy.

The whole question of the black hole entropy expression attributed to Bekenstein and Hawking has been dealt with in an earlier chapter. However, if the criticism discussed in that chapter has been around for as long as suggested there, why has it never really surfaced in the public domain and why has no answer been put forward? After all, the claim outlined is that the said expression is incorrect and could lead to violation of the Second Law of Thermodynamics – a possible consequence which, in normal circumstances, would be expected to create an immediate public furore!

One of the first, if not the first, criticisms of the said entropy expression appeared as a letter in the journal *Classical and Quantum Gravity* in 1988[*]. This letter was accepted for publication with no problems whatsoever for the authors. However, being a letter, it was obviously intended that detail of the criticism would be submitted at a later date, as is the usual practice in such situations. The follow-up explanatory article was duly written and submitted, but was rejected by the journal. No real challenge to this editorial decision was allowed. It was pointed out to the editor that the rejection implied that the original letter was incorrect and that, therefore, an article pointing out that incorrect information had been published in the journal should be produced. It was also commented that the authors of the original submission would obviously have automatic right of reply to such an article. The result was silence!

Over the years, several articles and books[†] have appeared in which the validity of the Bekenstein-Hawking expression for the entropy of a black hole has been queried. Not one has ever been criticised, - at least, not openly. All have been greeted with silence! It is not without interest to note that many more such articles have been rejected for publication with the

---

[*] B.H.Lavenda & J.Dunning-Davies, 1988, Classical & Quantum Gravity, 5, L149

[†] B.H.Lavenda, 1995, Thermodynamics of Extremes, (Albion Publishing, Chichester) and references cited there.

reason being advanced on more than one occasion that, although the referee could find nothing wrong with the article, the end result disagreed with Hawking and so must be wrong! Any truly open-minded person must see immediately the totally ludicrous and unsatisfactory nature of such happenings. However, the criticism will not go away and so, in the end, the tactic of silence will, hopefully, fail. Although it must be admitted that there is, as yet, little sign of a change as far as the thermodynamics of black holes, as attributed to Hawking, is concerned. A typical example illustrating this is provided in the recently published book '*Into the Cool*' by Eric Schneider and Dorion Sagan (University of Chicago Press, 2005). In this book, which is really devoted to attempting to apply thermodynamics – especially the second law – to problems in biology, the authors openly talk of the physics developed by Hawking to describe high-entropy black holes and go on to note that "Hawking's theory of the thermodynamics of black holes provides hope that the heat death of the universe imagined by the Victorians may never come to fruition". They speculate further that, if Hawking's ideas are correct, 'our descendants may rely upon black holes rather than stars for their energy needs'. This is all quite interesting to read but how accurate is it? In all probability, the statements to which reference has just been made are not too accurate at all; in truth, only time will tell. However, the worry is that these notions have been quoted in such an authoritative way in the year 2005. Concerns and criticisms of the thermodynamics of black holes have been clearly voiced, with no audible reply, for a long

Conventional Wisdom – Some Modern Case Studies

time now. Indeed, the same is true, as has been seen earlier, of the entire modern theory surrounding black holes. While few, if any, are saying such objects definitely do not exist, many are wondering over their precise nature if they are positively identified – as explained earlier, the theory based on the so-called Schwarzschild singularity certainly cannot be valid. It is not intended to review this book by Schneider and Sagan or even to comment on it overall, but merely to note that, even in 2005 after so many qualms have been raised in a wide range of publications, the popular view on the whole topic of black holes appears to have been accepted virtually without question. This again serves to raise the spectre of 'conventional wisdom'. It is undoubtedly the case that some knowledge becomes so well and so widely accepted that it is regarded as established fact. This would have been the situation with Newtonian mechanics. However, Einstein successfully challenged the all-embracing nature of that theory and, although doubts are still harboured concerning the validity of relativity, that challenge should be seen as a beacon in the history of science indicating that nothing, absolutely nothing, in science may be regarded as completely sacrosanct! If this simple truth is forgotten or discarded, progress in science – in *all* areas of science – will grind to a halt eventually.

This brief history of events surrounding the Bekenstein-Hawking black hole entropy expression brings back to mind the case of the paper by Albrecht and Magueijo on speeds faster than that of light, as discussed in Chapter One. The claim was made, at one point, that Moffat should have persevered with his articles to have them published in more 'prestigious' journals than those in which they finally did appear. However, most people's experience is that, if a paper is rejected, after one appeal correspondence is terminated by the journal's editor or editorial board. This was certainly the case where papers criticising the black hole entropy expression were concerned; discussion was stopped quite abruptly and with no truly satisfactory explanation of decisions being offered. In fact, on at least one occasion, when an article announcing the positive identification of a black hole appeared in the journal *Nature*, publication of criticism of that announcement on the grounds that the object in question did not satisfy the basic criterion for the ratio of its mass to its radius was refused, even though the authors of the original article agreed in writing that the criticism was valid, although they still felt they had identified a black hole positively! Once again, the question of how Albrecht and Magueijo managed to have their article published in the *Physical Review*, after apparent rejection, is raised. Appeals against original editorial decisions are only rarely listened to and even more rarely is the original decision reversed, but Magueijo's[*] history of events would seem to imply that he and Albrecht had quite a task in persuading the *Physical Review* to accept their article; - at least, this is certainly the impression given in the quoted reference.

---

[*] J.Magueijo, 2003, Faster than the Speed of Light, (William Heinemann, London)

It would be illuminating for science as a whole if the whole truth behind this particular story was made public.

2. The Tsallis Entropy.

Generally speaking, as far as physicists working in the broad area of statistical mechanics are concerned, the evaluation of the entropy of a system has depended on the use of the expression due to Boltzmann

$$S = k \ln W.$$

This expression has been used with great success for many years and is, in fact, carved on Boltzmann's tombstone. Suddenly, in 1988, a short paper appeared in the *Journal of Statistical Physics*[*], entitled '*Possible Generalization of Boltzmann-Gibbs Statistics*'. After a mere ten lines of introduction, the following formula was suggested as a generalisation of the above entropy expression

$$S_q \equiv k \frac{1 - \sum_{i=1}^{W} p_i^q}{q-1}.$$

As is immediately obvious, this expression is hardly of a type that one can imagine issuing forth as a product of pure thought for an immediate generalisation of Boltzmann's earlier one. However, quite soon, a great number of articles appeared using this formula to

---

[*] C.Tsallis, 1988, J. Stat. Phys. 52, 479

investigate an extremely wide variety of problems. Some referees did query the originality of the above expression, citing the book by Aczél and Daróczy, *On Measures of Information and their Characterizations*[*]. However, little notice was taken of these queries and the papers continued to appear regularly with no reference to earlier work. Over the years, an enormous body of literature based on this 'new' entropy expression accumulated. At the same time, however, it became apparent that the origins of the formula went back much further than the above mentioned book. It emerged that the formula had been well-known to information scientists for a great many years and certainly going back as far as a 1967 article by Havrda and Charvát[†], a fact that was finally acknowledged in 1995[‡]. However, a specialised form of this entropy has been found to date from 1912[§] and many useful results pertaining to it appear in the well-known book *Inequalities*[**] which was published originally in 1934.

In what amounts to a eulogy, which appeared in 2003[††], it was claimed that the idea of generalising the entropy and Boltmann-Gibbs statistical mechanics came

---

[*] J.Aczél & Z.Daróczy, 1975, On Measures of Information and their Characterizations, (Academic Press, New York)
[†] J.Havrda & F.Charvát, 1967, Kybernetika (Prague), 3, 30
[‡] C.Tsallis, 1995, Solitons & Fractals, 6, 539
[§] C.Gini, Studi Economico-Giurdici della Facoltá di Giurisprudenza dell'Universitá di Cagliari AIII, parte II
[**] G. Hardy, J. E. Littlewood & G. Pólya;1989, Inequalities, (Cambridge Univ. Press, Cambridge)
[††] R.Graham, 2003, Santa Fe Institute Bulletin, 15, no. 2 (www.santafe.edu/sfi/publications/Bulletins/bulletinFall00/features/tsallis.html)

Conventional Wisdom – Some Modern Case Studies

to Tsallis during a coffee break at a workshop in Mexico City in 1985. Without wishing to appear in any way blasphemous, at this point the question of whether or not this was meant to indicate a divine revelation springs to mind. However, it took a further three years before this original idea came to publication. It is also claimed later in the same article that Tsallis explained his ideas to a reporter and wrote material down so that the power of the seemingly simple approach became apparent. None of this appeared in the 1988 paper; just a bald statement of the new expression – no derivation, no indication even of what this 'simple approach' constituted. In many ways, this is an unsatisfactory situation scientifically but, fundamentally, there is nothing terribly wrong with this story so far. Someone working in the field of statistical mechanics could be unaware of results in such a separate field as information theory very easily. However, where this story links up with that which has preceded it is through the all-pervading notion of 'conventional wisdom'. This idea was so outside the norm of statistical mechanics that, in some ways, it is surprising that it was ever published, especially since no indication of a derivation was included; for that, though, the editor of the journal concerned is deserving of some credit. However, far more surprising is the fact that it became accepted so widely and so quickly. On top of this, comes the query as to why questions raised about its total originality seem to have been ignored. It is easy to understand this situation existing in more recent times since, by the year 2000, the expression had become so well-established, it was probably regarded as coming under the protective umbrella of 'conventional wisdom', but it is not so easy, indeed it is well-nigh impossible, to understand how earlier queries came to nothing. The worrying fact remains that criticism of the origins of this expression, as well as of the expression itself and deductions based on it, are difficult to find in what are regarded normally as the 'prestigious' journals. The pages of these journals remain closed to those who raise queries about both the so-called 'Tsallis' entropy and work based on it. Why? In truth, science should thrive on informed criticism; indeed, is that not one of the means by which true science is seen to progress?

The total volume of scientific literature has increased enormously in recent years. New journals seem to be appearing with almost monotonous regularity and many of the older more established ones are expanding so much as to require considerable physical effort when viewing the paper versions. Few, if any, may be expected to keep fully up-to-date with this manic proliferation of information. Hence, once again, the truth of the claim that 'no man is an island' is forced on each and every one of us. All scientists desperately need the help of others to ensure that older references especially are not missed in the publication morass. One task of a referee is to spot important references that have been missed, albeit inadvertently. It is then the bounden duty of *all* journals – editors and editorial boards – to ensure that due notice is taken of this extra information, both when making a final decision on accepting or rejecting an article and, possibly more importantly, in the successful cases where the article is accepted for publication. The older references cannot simply be ignored as a matter of personal convenience. Undoubtedly, this is the procedure which should have been applied in the case of the original publications involving the 'Tsallis' entropy.

It might reasonably be commented that the general situation is likely to become much worse in the not too distant future. Many volumes of journals are now stored on computer networks but, often, these go back only to about 1981. Are we to be faced by a situation when only post-1981 publications are recognised? If so, the pressure to publish induced by such activities as Research Assessment Exercises will surely induce people to publish afresh results which have actually been in the public domain for a great many years. This is an area which will undoubtedly require extremely careful monitoring in the years ahead.

3. *The Inflationary Scenario.*

Yet another case where 'conventional wisdom' seems to have played a part is that surrounding the notion of the inflationary scenario which, as discussed earlier, was introduced basically to shore up the 'big bang' theory. In order to explain some of the problems associated with the 'big bang', in 1981 Guth[*] originally introduced the idea of inflation into cosmology. This theory certainly seemed to solve some of the problems facing that theory and has been elaborated since.

---

[*] A. H. Guth, 1981, Phys. Rev. D. 23, 347

However, in the original article, in order to resolve the problems faced by hot big bang cosmology, Guth released the assumption of adiabaticity. Unfortunately, releasing this assumption proves incompatible with the Einstein equations, as was noted earlier. By the time, this was noted – some ten years later – inflation had become well-established and a great many articles had been written about it and even more had made use of it. Consequently, it was extremely difficult to gain acceptance of the query for publication. Publication was achieved eventually[*], but only after considerable problems – not one of which claimed the criticism actually incorrect. In retrospect, it would seem that, after a lapse of ten years, inflation had become established as part of 'conventional wisdom' and, as such, might be regarded, therefore, as almost sacrosanct. In this particular case, the fact that inflationary theory was supporting the 'big bang' theory, - in fact was playing and still does play a major role in propping up that theory to the exclusion of alternatives, - would probably have given additional impetus for its inclusion under the 'conventional wisdom' umbrella.

4. String Theory.

Very recently, a truly fascinating, highly informative volume concerned with the topic of string theory has appeared in the bookshops. The book, *Not Even Wrong*

---

[*] B. H. Lavenda and J. Dunning-Davies, 1992, Found. Phys. Lett. 5, 191

Conventional Wisdom – Some Modern Case Studies

by Peter Woit[*] examines in detail the history of this branch of physics which supposedly deals with the most up-to-date theory of elementary particles. To the uninitiated, the whole theory seems highly mathematical and abstract; certainly not immediately appearing as a branch of physics dealing with what are thought to be the fundamental building blocks of nature. Here, in my view, Peter Woit succeeds brilliantly in describing a highly mathematical subject in a way which enables the non-mathematically inclined to obtain a good understanding and appreciation of what has happened, is happening, and possibly most importantly what is likely to happen in the future in this technical field. The latter point is of particular importance when the impact on the non-scientific public is considered. As with all areas of research, it is the general public which foots the bill in the end and so, in these days of increased efficiency and effectiveness of communication, it seems only correct that that public is made fully aware of the manner in which its funds are being utilised. In the area of elementary particles, although much effort is put into theoretical considerations, any such considerations have to be verified in practical situations. This necessitates the building of bigger and bigger particle accelerators at truly enormous cost. Peter Woit describes the entire process – both the theoretical modelling and the attempts at practical verification – in graphic detail. He begins by describing the mathematical deliberations and does so purely by means of the English language with

---

[*] P. Woit; 2006, Not Even Wrong, (Jonathan Cape, London)

Exploding A Myth

not a mathematical symbol in sight. He then discusses the problems faced by the experimentalists, noting particularly the costs involved. He concludes by summing up the position in which the physics community now finds itself. He notes that the area has helped to spawn great advances in some areas of pure mathematics but, while admitting the honest intentions of those involved, doubts that in over twenty-five years of tremendous effort anything of real note has been achieved as far as elementary particle theory is concerned. Here is a man who has worked both as a physicist and, latterly, as a mathematician in the field but who is able to say that, with all the endeavour, string theory has yet to produce even one testable prediction.

The above may be an interesting point to note but where does it fit in with the general thesis under consideration here of the influence of so-called 'conventional wisdom' on scientific research? To gain a full appreciation of the link, it would be sensible to read Peter Woit's book. However, with his inside knowledge of the American university education system, he is well-placed to be able to point to the influence exerted by string theorists in many areas of that system. To use his own word, he feels a 'mafia – like' influence is in existence. Tremendous pressure seems to be being brought to bear in many areas of education to support the continuation of this seemingly pointless investigation of string theory, and its more modern extension, M – theory, even though little of physical note is appearing. When the costs involved with the

Conventional Wisdom – Some Modern Case Studies

construction of the newer particle accelerators are considered, thoughts must surely switch to such projects as the running of the experiments necessary to check out Santilli's theories concerning the safe disposal of nuclear waste, as mentioned in Chapter Five. Here the cost of the required experiments would probably be of the order of a million pounds. A lot of money; yes! However, if the experiments proved successful, the benefits for mankind would be out of all proportion to the money spent. If unsuccessful, a lot of money would have been wasted, but nothing in comparison with the sums spent on the construction of huge new particle accelerators; such a machine would cost millions of pounds – indeed a new detector alone could cost as much as, or possibly more than, a million and a half pounds! The point here, however, is that those deeply involved with string theory research are in a position to exert enormous influence and so, this is another area which has come under the umbrella of 'conventional wisdom'. Its influence has been strong for many years, possibly because of the association of many of its leading practitioners with other influential areas of physics, such as 'Big Bang' theory and black holes. It was, in fact, many years ago when heterotic strings were felt to be important that it was noted that the claimed expression for the entropy of such strings violated the Second Law of Thermodynamics[*]. However, as with the problems with the entropy expression for black holes, the point proved difficult to make since it conflicted with 'conventional wisdom'

---

[*] J. Dunning-Davies, 1999, Hadronic Journal, 22, 117

Exploding A Myth

and, when it did appear in print, it was studiously ignored – again paralleling the black hole entropy expression situation. Unfortunately, in one respect at least, the conduct of scientific research accurately mirrors everyday life and that is, if a tricky problem arises, the best policy may be to ignore it because, if you do, that problem will slowly drift away in the vast majority of cases. This cannot be seen as a surprising situation since, after all, all scientists are human beings like everyone else; it's simply that they have a different occupation.

In the introduction to his book, Peter Woit explains that one of the main appeals of science to him was that it "involved a notion of truth not based on appeal to authority." He notes further that "judgements about scientific truth are supposed to be based on the logical consistency of arguments and the evidence of experiment, not on the eminence of those claiming to know the truth". This statement accurately mirrors the sentiments expressed throughout this book and is another way of calling into question the effects of 'conventional wisdom' on scientific research itself and, incidentally, on the distribution of funds to allow various scientific projects to be pursued while others are not.

These varied incidents raise, once again, questions over the nature of this 'conventional wisdom' phenomenon which seems so powerful in science today. The case of Waterston indicates that it is not anything completely new, but that is not a valid justification for it and certainly doesn't make it correct. Further, one very worrying aspect of this is that all the examples discussed here have been related to areas of physics and, while physics may, and indeed does, exert tremendous influence in all fields of science, to the extent that it may truly be thought to be at the centre of all scientific endeavour, it must follow that this pervasive cancer of 'conventional wisdom' exists in other areas of science also. This could mean that its influence exists in medicine, for example, and here, if its influence caused the incorrect withholding of progress, the death of patients could easily be an end result. It is possibly this realisation that indicates the truly evil nature of this notion. 'Conventional wisdom' may be an extremely convenient artifice for protecting the status quo and, therefore, the positions of those who have built their reputations on foundations that are less than solid but, unfortunately, it is a situation wide open to abuse that could have tragic consequences and so, it is one which must be abandoned completely and immediately.

Chapter Seven

Some Final Thoughts

It might seem that a slightly ambivalent attitude has been displayed throughout towards Einstein. Undoubtedly serious queries have been voiced concerning his theories of relativity. However, his own concerns about the range of validity of quantum mechanics have been raised also. The ignoring of his published views on the singularity leading to the idea of a relativistic black hole has been mentioned. However, what of Einstein the man? It is well-known that he was deeply disturbed at the impact on humanity of many scientific discoveries and advances that had occurred during his lifetime. His concerns over, for example, nuclear weapons have been well documented. However, in some ways, his wonderful humanity and compassion are exemplified by what is probably a little known true story. The American author John Gunther, well known for his books such as *Inside Asia*, *Inside U.S.A.*, and *Inside Russia Today*, also wrote a short volume entitled *Death be not Proud*, the title being taken from a poem by John Donne. This extremely moving memoir catalogues the final period in the life of his son, John

Some Final Thoughts

Gunther Jr., who died of a brain tumour aged just seventeen. He had been a very intelligent young man, who was extremely interested in science amongst other things. While seriously ill, he had been thinking about physics and eventually wrote to Einstein. His father enclosed a covering letter. Einstein himself found the time to reply in person. It is recorded how much this meant to all the Gunther family at that time of deep anguish. This little story is likely to pass unnoticed but, as indicated earlier, is surely a measure of Einstein the man. Incidentally, the letter John Gunther Jr. wrote was seen by another eminent physicist of the time who expressed amazement that a young man of seventeen should have even been aware of the problem – essentially the notion of a unified field theory - he was raising, never mind raise it in a sensible, rational way in a letter.

As for Einstein, it has to be concluded that he was a truly great scientist, as well as being a kind, compassionate man. No doubt he had his faults. Who doesn't? However, the anecdote concerning the Gunther family leaves little doubt that he had genuine compassion for others. This was not a case of writing to the President of the United States of America on issues concerning nuclear weapons, an occurrence which was bound to produce attendant publicity – not that he necessarily wanted such publicity; this was an eminent man taking the time to write privately to a desperately ill young man and doing so out of simple human compassion. Surely no other conclusion may be reached? His theories of relativity were great

Exploding A Myth

intellectual achievements and, even if there are doubts about the range of their validity, that assessment of those theories must remain. The theories of relativity have become so well-known in popular science and to the populace at large that his other achievements are frequently overlooked. His work on Brownian Motion alone, to which reference was made earlier, is in itself cause for regarding him as at least one of the greatest scientific figures of the twentieth century. Although all physicists know of the so-called Bose-Einstein statistics, this is another area of his work which is often neglected. In fact, it was only towards the end of the last century that much of his pioneering work on heat capacity and the Bose-Einstein condensate began to produce results of real importance in the world of physics. All this is, of course, leaving until the end the work on the photoelectric effect, for which he received a Nobel Prize. To many, the truly surprising thing about Einstein is that, as a Nobel Prize winner, he is remembered by most for work other than that for which he was awarded the prize. Many genuinely believe he should have been awarded a second prize for the work on relativity; - in fact, it is quite possible that many non-scientists believe he was awarded the prize for his work on relativity. From this point in time, noting the public adulation for that work, it is interesting to wonder why, in fact, that second prize was never awarded. It seems that, on occasions, funny are the ways of the scientific establishment. However, reflecting on what has been the overall linking subject of these pages and of the ways in which Einstein's name has occurred, one big query emerges and that is, where did Einstein *really*

Some Final Thoughts

stand with the scientific establishment? There can be no doubt that there are occasions where his work is regarded as being a crucial integral component of accepted 'conventional wisdom'. This is very obviously true of the basic theories of relativity, which are defended by influential members of the scientific establishment with a zeal of a kind normally associated with extreme religious behaviour. Anyone who questions the authenticity of these theories is immediately condemned as a pariah in scientific society and, if at all possible, cast into the proverbial outer darkness. However, why is it then that his pronouncements on matters closely associated with these basic theories are ignored? Why is it that, on so many occasions, the general public is led to believe this man is well-nigh infallible on matters associated with several areas of physics but still his published work in some of those actual areas is conveniently ignored, or forgotten, when it seems to suit certain vested interests? Here obvious reference is indicated to his very clear pronouncement on the question of the so-called Schwarzschild singularity and the existence of black holes such as many claim are predicted by general relativity. Also, it must never be forgotten that he always retained grave doubts over the lack of deterministic character of quantum mechanics; this by the man who has given science Bose-Einstein statistics, the theory of the photo-electric effect, a detailed theory of heat capacity and is the founding father of the theory behind Brownian motion. Evidently, 'conventional wisdom' must even be allowed to transcend genuine scientific greatness on occasions when it suits some of lesser ability, but great pseudo-political power.

It is interesting, and not a little ironic, to note that there are many who have wondered over the years why Einstein didn't receive a second Nobel Prize, one for his fundamental work on relativity or possibly for his fundamental work on Brownian motion. In what has gone before, mention has been made of another great contributor to modern scientific knowledge who failed to be awarded even one Nobel Prize; Sir Fred Hoyle! When one reflects back on the scientific achievements of recent Nobel Prize winners in physics, no-one can doubt the correctness of awarding such a prize, in 1983, to Fowler for his studies of the nuclear reactions of importance in the formation of the chemical elements in the universe. However, the fundamental paper associated with these studies is the classic one from 1957 by E. M. Burbidge, G. R. Burbidge, W. A. Fowler and F. Hoyle (*Rev. Mod. Phys.* **29**, 547). Also, the actual origins of this work go back to initial studies begun by Hoyle in the mid 1940's. In fact, by about 1946, Hoyle himself had discovered the series of chain reactions which would build the elements from carbon to iron in equilibrium nucleosynthesis. It doesn't seem that these stories regarding the beginnings of the theory are in any way apocryphal and so, if that is indeed the case, the question of why Hoyle was never honoured with a Nobel Prize – or at least a share in one – has to be raised. The fairly recent death of Fred Hoyle has spawned a number of books about him. Hoyle's lack of a Nobel Prize has, quite naturally, been a topic for

Some Final Thoughts

inclusion and it is interesting, though rather sad, to read that some feel his own nature and temperament might have contributed, or even led, to his not being awarded such a prize. Whatever the truth may be, and probably no-one will ever know, it might be felt that it is more the Nobel Prize's loss than Hoyle's that he received no such award. However, be that as it may, it is disturbing that anyone might even be led to think that a person's temperament could be a factor in the award of a scientific prize. This seems another incidence of a form of 'conventional wisdom' rearing its ugly head to the detriment of science. Was Hoyle really such an 'enfant terrible' in the world of science? It is definitely for others to answer that question, but it is surely information which can have absolutely nothing to do with a person's ability as a scientist or, indeed, their achievements in science. Hence, it is, as noted before, not a factor which should be allowed to enter the discussions surrounding the award of any scientific prize. As a final point concerning Hoyle, which seems to link him to the perceived problem of 'conventional wisdom', it is of interest to note that, following lectures broadcast by the B.B.C. in 1950, he combined the material covered into a short volume entitled *The Nature of the Universe* (Basil Blackwell, Oxford, 1950). In a 1960 version of this book, Hoyle speculates that he has assumed throughout that, in the future, progress will be made, but it is possible that, as happened to Greek astronomy after Hipparchus, it might go backward. He comments that, in taking this view he is "not thinking about an atomic war destroying civilisation, but about the increasing tendency to rivet scientific enquiry in fetters". Once again, it seems, a warning concerning the dangers of 'conventional wisdom' to progress in science has been allowed to pass unheeded.

Considering the earlier discussion in Chapter Five of the work of Ruggero Santilli, it is obviously in order to discuss his contributions to the physical sciences at this juncture. Quite obviously, the jury is still out considering its verdict. The body of work to be examined is quite enormous and all of it needs to be considered fully and with open minds, - minds totally unencumbered by the constraints which could, and by many would, be imposed by 'conventional wisdom'. 'Conventional wisdom' simply must not be allowed to influence this verdict in any way at all; the verdict must be delivered on the basis of genuine scientific facts, both theoretical and experimental. The suggestions under discussion are very basic, since they go all the way back to constructing several new forms of mathematics. There is little new in this; basically, Einstein made use of what might in his time have been called 'new mathematics' when he introduced the work of Riemann, Ricci and Bianchi to the world of physics through his general theory of relativity. However, the introduction does introduce unusual problems into the evaluation process because it involves those charged with the task entering totally new, uncharted waters and that can prove a daunting – even frightening – prospect. Nevertheless, it is a nettle that must be grasped and, in the case in question, the possible benefits for mankind make the fair and intellectually honest completion of that task all the more urgent. There can surely be little

Some Final Thoughts

doubt that, if the verdict of the jury is positive, Santilli also should be considered for the award of a Nobel Prize.

In the three examples cited above, the first two people concerned have been truly great men of science by any standards and the discussion has centred around whether or not they should have been awarded Nobel Prizes; those prizes regarded, rightly or wrongly as the greatest of academic accolades. In one of the first two cases, the discussion is concerned, of course, with whether a second such prize should have been awarded. In the third case, the person's academic eminence is not rigorously established as yet and the discussion revolves around the question of whether the award of such a prize should be considered in the event of that eminence being recognised. Unfortunately, 'conventional wisdom' doesn't restrict its attention to simply one category of practitioner, or even to one subject. The examples considered here have all been in the physical sciences. It is impossible to imagine that the 'virus' doesn't allow its tentacles to infiltrate other areas of academic endeavour; indeed to pervade any activity where human jealousy and envy can play a part. Tragically, as indicated earlier, this would have to include medicine and then the possible effects not only retard development, but could put lives at risk. This latter point is possibly the most worrying. Is the general area of medicine affected by this problem? Only practitioners can say but, as noted already, given human nature being what it is, it is impossible to imagine any

Exploding A Myth

area of intellectual endeavour remaining impervious to its obvious attractions.

Further, the 'virus' doesn't restrict its attention to the academically most eminent either. It seems to pervade all levels of academia in a wide variety of ways and may even be a factor in determining a person's actual standing in the academic community. If a junior person inadvertently treads on the toes of someone who has been elevated – rightly or wrongly - to a position of prominence, 'conventional wisdom' may well be invoked, directly or indirectly, to effectively stifle that junior person's career. This echoes the scenario discussed so clearly by Lord Rayleigh when he was introducing the long-neglected article on the kinetic theory of gases by J. J. Waterston. If the young researcher follows Lord Rayleigh's advice, he will be drawn into the seductive influence of 'conventional wisdom'. He will produce non-controversial material to establish his reputation but, by the time he has obtained that desired reputation, he will have been seduced into membership of that same establishment which controls 'conventional wisdom'. It is indeed a vicious circle from which few escape. For all the clear warnings embedded in the case of Waterston, the position has certainly not changed since those days at the end of the nineteenth century; indeed, many might feel the position has worsened. In any event, knowledge of what happened to Waterston and the ensuing consequences for science as a whole seem to have had little, or no, effect on the inner workings of the scientific establishment. Whether or not this final thought is

Some Final Thoughts

completely valid, however, is of little importance. The fact remains that within physics certainly, and probably within all other areas of science, the cancerous effects of 'conventional wisdom' are witnessed every day by scientists. Not all will admit to this, but then, as with everything, someone benefits, and so it is definitely not in their interest to expose it. However, exposed it must be! Does anyone know what possible benefits for mankind are lying hidden in some dusty archive because 'conventional wisdom' effectively decrees they must not see the light of day? Is it possible to dispose of nuclear waste safely in-house, with no need for dangerous transportation or incredibly lengthy burial? Are there other sources of clean energy available for the use of mankind if only work was allowed to open-mindedly search for them? Could there be cures for serious diseases available if only some unorthodox views were not kept hidden? At the present time, these are simply questions that may be posed with little real hope of complete and honest answers. Considering the problems facing mankind now and in the future, these problems must be faced openly and, to do that effectively, the power and seductive influence of 'conventional wisdom' must be destroyed. No one person, in any field, may be regarded as omnipotent; no theory may ever be regarded as offering an ultimate, complete answer to any particular problem. In true science, **all** theories and ideas must remain open to scrutiny – however well-established they may be!

A further major cause for concern relates to modern day scientific education. There is no doubt that many

Exploding A Myth

young people are taught about various theories in science as if they were actually indisputable fact rather than just mere theories. Apart from anything else, this approach to teaching has a tendency to remove any inclination to think. So often these days, the will and ability to think for oneself are discouraged. Classics as an academic subject is in decline; few study it at school; even fewer go on to study it at university. However, Classics might reasonably be termed a *true* university discipline. It is often said that Latin and Ancient Greek are dead languages and, therefore, of no use to anyone. As such, many seem to feel it a waste of time studying them. In truth, everyone can learn a tremendous amount from studying these subjects, both the languages and the literature, but, in some ways, the most important thing about studying Classics is that people who do so are taught to think! The structure of these two languages is such that studying them imposes a discipline on the student, similar to that which may be imposed by the study of mathematics. At the end of an undergraduate degree course, a Classics student might not have absorbed numerous facts which will be useful immediately in everyday life, but their mind will have been trained to think to such an extent that they should be able to apply that ability to considering problems in a wide variety of fields. However, the ability to think also implies an ability, even a desire, to question and it would seem that that is not something the adherents to 'conventional wisdom' would want. All too often these days, students are being taught material as if it was a statement of the truth, the whole truth and nothing but the truth. In fact, what they are being taught are theories

Some Final Thoughts

and, in some ways more importantly, the theories in vogue at this moment in time. As claimed earlier, 'conventional wisdom' can be something which serves to stifle original thought and, therefore, hinder true progress. If the up and coming generation is to be taught this material as if it is sacrosanct, the ramifications could be even more far reaching and disastrous. Obviously, the effects will be greater in subjects such as astronomy and cosmology because so many of the theories in those fields cannot be examined easily in a way which might lead to readily verifiable confirmation. If the lifetime of a star, for example, is under consideration, the value will be so large that there is no way the scientist involved can even begin to think of viewing both the birth and death of that object. This is so true of virtually all examples in those two disciplines. Hence, they are both composed of theories which admit only indirect confirmation, or otherwise. As mentioned previously, it is absolutely vital for science in general that this aspect is not forgotten. Today, money for scientific research is not so readily available as it once was. Therefore, when projects are considered for financing, they must be considered purely on the basis of their academic worth; peripheral issues such as the dictats of 'conventional wisdom' must be allowed no say in such matters. However, the financing of scientific projects is not the most important aspect here; obtaining true answers to scientific questions far outweighs it in every way. Whether it be cosmology or astronomy, particle physics or magnetics, biophysics or medicine, the search for scientific truth must never be hampered by unnatural, man-made barriers such as those imposed by 'conventional wisdom' and all students must be trained to think for themselves, to possess open minds in which novel ideas may be able to flourish without artificially imposed constraints.

Epilogue

The original intention in writing this short book was to draw attention to problems associated with the popular views of some well publicised scientific theories; notably the theories of relativity, the notion of black holes, and the Big Bang theory. Throughout, the importance to public perception of these matters by what is termed 'conventional wisdom' became more and more apparent. This took on yet more significance when the work of Ruggero Santilli came to be considered. It was this that eventually caused the emphasis to be placed on a consideration of the role of 'conventional wisdom' in modern science. The end result of that examination is not to instil confidence in the future development of world science. The case of Santilli alone – whether his theories prove correct or not – serves to convince that the minds of the ruling élite of world science are not entirely open. This must be having a detrimental effect on the training and development of the young scientists on whom the future depends. Minds must become more open; theories must not ever be allowed to become unchallengeable; 'conventional wisdom' must be modified so as to be less pervasive in this whole area of human endeavour on which our very existence depends.

It might, in conclusion, be remembered that, in his book *Science and Hypothesis*, (Dover, 1952), the great Henri Poincaré states that

Exploding A Myth

"To doubt everything or to believe everything are two equally convenient solutions; both dispense with the necessity of reflection".

The scientific community throughout the world might do well to take this thought to heart!

Index

Abraham, M., 57
Abrams, L.S.,141
A Brief History of Time, 45
Aczél, J.,228
Adler, R.,132
aether, 33, 50, 54, 55, 56, 58, 65, 66, 69, 72, 77, 159
Albrecht, A.,24, 26, 27, 30, 31, 33, 35, 37, 39, 40, 41, 43, 87, 144, 225
Alpher, R.,106, 111, 113, 114
Antoci, S.,131, 134, 139
Arp, H.,93, 95, 96, 98, 123, 125
Aspden, H.,74, 75, 77, 79, 80
Barrow, J.,33
Bazin, M., 132
Becker, R.,57
Bekenstein, J.,162, 168, 222, 223, 225
Bell, J.,150
big bang, 21, 23, 26, 86, 87, 88, 93, 98, 99, 100, 101, 104, 105, 106, 110, 117, 119, 122, 124, 127
black hole, 132, 133, 138, 142, 153, 160, 163, 168, 176
Bohm, D.,183
Bohr, N.,184, 185
Boltzmann, L.,227
Bondi, H.,99, 101, 111
Borghi, C.,208
Born, M.,110, 123, 185
Brillouin, M.,141
Brush, S.G.,61
Burbidge, G.,98, 99, 118
Cameron, G.H.,110
Carathéodory, C.,170
Castellano, V.,187
Chadwick, J.,205
Chandrasekhar, S.,144, 148
Charvát, F.,228
Cole, G.H.A.,29, 56, 187
Cowling, T.G.,58
Crab Pulsar, 151
Crothers, S.J.,141
Daróczy, Z.,228
de Broglie, L.,185
Dingle, H.,68, 69, 70, 72

Dirac, P.A.M.,134, 136, 185
Dunning-Davies, J.,29, 39, 40, 56, 143, 187, 223
Earnshaw, S.,78, 79
Eddington, A., 72, 107, 108, 109, 110
Einstein, A.,23, 27, 28, 51, 67, 69, 71, 72, 74, 81, 83, 89, 116, 131, 133, 134, 136, 137, 138, 143, 161, 175, 178, 181, 184, 185, 195, 198, 238, 239, 240, 244
Einstein's Biggest Blunder, 23, 28, 41, 43
Equinox, 23
Evans, R.F.,187
Faster than the speed of light, 41
Fellgett, P.B.,128
Friedmann, A.,89
Gamow, G.,89, 106, 110, 111, 113
General Theory of Relativity, 132, 134, 175
Genesis, 101, 104
Gerber, P.,80, 81
Gibbs, J.W.,173
Gibbs-Duhem equation, 164
Gini, C.,228

Gold, T.,99, 101, 106, 111, 150
Graham, R.,228
Guillaume, C-E.,109, 110
Gunther, J.,238
Guth, A.,35, 36, 120, 231
Hadronic Chemistry, 189
hadronic mechanics, 20, 21, 184, 195, 203, 208, 215
Hamilton, W.R., 83, 201
Hanbury-Brown, R.,118
Hardy, G.,228
Havrda, J.,228
Hawking, S.W.,45, 120, 162, 168, 170, 174, 222, 223, 225
Heisenberg, W.,183, 185
Herman, R.,111, 113, 114
Hertz, H.,59, 61, 83
Hewish, A.,150
Hilbert, D.,142
Hoyle, F.,98, 99, 101, 104, 105, 111, 113, 117, 118, 242, 243
Hubble, E.,89, 92, 108
Illert, C.,220
Jeans, J.,77, 78
Johansen, S.,106
Kaufmann, 75
Lagrange, J-L.,83, 201
Landé, A.,183
Lavenda, B.H.,36, 81, 82, 83, 171, 223
Lemaître, G.,89
Lie, M.S.,195

Index

Linde, A.D.,120
Littlewood, J.E.,228
Lodge, O.,72
Loinger, A.,131, 134, 138, 139
Lorentz, H.A.,55, 58, 69, 71
Magueijo, J.,24, 25, 27, 30, 31, 32, 33, 35, 38, 39, 40, 41, 43, 44, 46, 49, 51, 52, 87, 144, 225, 226
Maxwell, J.C.,59, 61, 64, 65, 179
Maxwell equations, 57
McKellar, A.,108, 111, 116
Michell, J.,130, 174, 175
Michelson and Morley, 50 54, 55
microcanonical ensemble, 171, 173
Millikan, R.A.,110
Minkowski, H.,64
Moffat, J.,28, 32, 38, 42, 46, 52, 144, 226
Moss, I.,120
Narlikar, J.,45, 98, 99, 118
Nature, 47, 48, 76
Nernst, W.,110
neutron star, 144, 149, 155, 158
Newton, I.,76, 82, 87

Okun, L.B.,67
Oppenheimer, J.R.,89, 137
Pauli exclusion principle, 147
Penzias, A.A.,89, 107, 111, 116
Phipps, T.E.,59, 60
Planck, M.,163, 185
Podolsky, B.,195, 198
Poincaré, H.,55, 71, 252
Pólya, G.,228
Popper, K.,182, 183, 184
Prandtl, L.,49, 56, 66
Rayleigh, Lord.,61, 62, 76, 178, 179, 246
Rees, M.,127
Regener, E.,110
Ritz, W.,72
Robertson-Walker metric, 120
Rosen, N.,195, 198
Rutherford, E.,51, 205
Ryle, M.,118
Sagan, D.,224
Santilli, R.M.,184, 185, 186, 190, 193, 195, 196, 198, 204, 206, 207, 209, 210, 214, 215, 218, 220, 235, 244, 245, 251
Schiffer, M.,132
Schneider, E.D.,224
Schrödinger, E.,185

Schuster, A., 61
Schwarzschild, K., 131, 134, 135, 136, 139, 141, 143, 168, 175, 177, 225, 241
Scott, W.T., 79
Second Law of Thermodynamics, 163, 222, 235
Shipley, A., 61
Singh, S., 115
Snyder, H., 137
Soddy., F., 51
steady state, 100, 101, 107, 117, 118, 122, 124
Stokes, G., 49, 55, 56
'*The Primæval Universe*', 45

Thomson, J.J., 75, 76
Thornhill, C.K., 29, 31, 32, 34, 37, 39, 41, 43, 52, 56, 57, 58, 60, 63, 64, 66, 73, 77, 144
Tsallis, C., 227, 229, 230
Vela Pulsar, 151
Vigier, J-P., 123, 183
Wark, D., 31, 37, 39, 40
Waterston, J.J., 61, 62, 76, 178, 246
Welch, A., 127
Whetham, W.C.D., 75
white dwarf, 144, 148, 155, 158
Wilson, R.W., 89, 107, 111, 116
Woit, P., 233, 236